Marcelo Divino Ribeiro Pereira

Socio-economic interfaces of those affected by dams: Estreito-MA

Marcelo Divino Ribeiro Pereira

Socio-economic interfaces of those affected by dams: Estreito-MA

Socio-economic impacts and the compensation process

Imprint
Any brand names and product names mentioned in this book are subject to trademark, brand or patent protection and are trademarks or registered trademarks of their respective holders. The use of brand names, product names, common names, trade names, product descriptions etc. even without a particular marking in this work is in no way to be construed to mean that such names may be regarded as unrestricted in respect of trademark and brand protection legislation and could thus be used by anyone.

Cover image: www.ingimage.com

This book is a translation from the original published under ISBN 978-613-9-71984-6.

Publisher:
Sciencia Scripts
is a trademark of
Dodo Books Indian Ocean Ltd. and OmniScriptum S.R.L publishing group

120 High Road, East Finchley, London, N2 9ED, United Kingdom
Str. Armeneasca 28/1, office 1, Chisinau MD-2012, Republic of Moldova, Europe
Printed at: see last page
ISBN: 978-620-7-95066-9

ACKNOWLEDGEMENT

Firstly, I thank God for my health and the conviction that everything is possible when we believe in him and in our dreams.

I would like to thank my wife, Poliana Ramos dos Santos, for her support and motivation during this time of complete dedication to my studies. To my parents, Sabina Ribeiro de Oliveira, João Pereira de Oliveira (in memorian) and José Pereira da Silva (stepfather), and, by extension, to my brothers, sisters, nephews and nieces, for their encouragement and for believing that I could get there, no matter how difficult this journey was.

To CAPES, for the scholarship I was awarded to dedicate to this research. To my professor and advisor, Dr Antônio Pedroso José Neto, for his rich contribution over the two years of my master's degree, with countless suggestions that were crucial to my intellectual maturity in relation to this research and others that I will develop later.

To Professor Dr Waldecy Rodrigues, for the opportunity to discuss good and timely texts on economic development. Professor Dr Temis Gomes Parente, for the dialogue with good texts, from which good ideas emerged. Professor Dr Jean Nascimento, who, regardless of the context, is a person who transmits confidence and security in relation to university life; Professor Dr Mônica Aparecida, for the great contribution she made to this dissertation, when she enabled me to come into contact with her texts on public policies and qualitative research methods. I would also like to highlight the great contribution of Professor Bia, whose lectures are unforgettable, even though her time on the programme was very short-lived.

To all the colleagues I met on the programme, both in the 2010 and 2011 cohorts, because it was a lot of "suffering" to complete this master's degree.

My special thanks go to Lorrane (who gave me so many rides); Michele, for "suffering" together in the 2011 class; Bárbara, for clearing up some of my doubts about Terra View Geoprocessing; and Thiago, from Rio Grande do Sul, a soya planter in Buritirana, for his contribution to this research; to Jorge Sánches, who I consider to be one of my teachers in the programme, because through this *el muchacho I* learnt how to use quantitative methodology (multiple regression), which will be of great importance for the development of future research; valeu *brother* This is knowledge that I will carry with me throughout my life.

Wana and Lethicia for their efficient and effective work as secretaries for the students on the programme. Thanks!

SUMMARY

The construction of large hydroelectric projects in Brazil since the last quarter of the 20th century is directly related to its industrialisation process, which was impacted by the oil crisis of the 1970s, when the government decided to invest in other alternative energy sources, including the construction of large hydroelectric plants in practically all Brazilian regions. With the implementation of these large hydroelectric plants in the South and Northeast, at first, and studies being carried out for the hydraulic use of other river basins located mainly in the North of Brazil, the Movement of People Affected by Dams (MAB) emerged, also on a regional scale, made up of countless riverside communities that had been dispossessed of their livelihoods as a result of the construction of large hydroelectric projects. The main aim of this research is to describe and analyse the socio-economic impacts of the socio-political process of financial compensation for the riverside communities affected by the Estreito Hydroelectric Power Plant (HPP), one of the large projects sponsored by the Brazilian government. The Estreito hydroelectric plant, the focus of this case study, is located in the city of Estreito, in the south of the state of Maranhão. By applying questionnaires to characterise the socio-economic situation of the affected communities, we sought to understand the extent to which their social, political, economic and cultural representations may have influenced the compensation process. Through interpreting the data, we found that the relative position of the individuals in the communities analysed, based on the aforementioned economic, political, social and cultural capitals, did not influence the compensation, considering that the Estreito Energia Consortium (CESTE) opted for compensation based on the land title.

Key words: Hydroelectric power stations. Socio-economic impact. Financial compensation.

INTRODUCTION

The aim of this study is to analyse the socio-economic impacts and the socio-political process of compensation for the communities of fishermen, barraqueiros, barqueiros, vazanteiros, farmers and local traders affected by the Estreito Hydroelectric Power Plant (HPP), built on a stretch of the middle Tocantins, in the state of Maranhão.

The data collection analysis for this research took place between 2011 and 2012 in the city of Estreito, where the power plant of the same name is located, which is justified by the fact that it is situated in one of the narrow "spans" of the Tocantins River, between the cities of Aguiamópolis (TO) and Estreito (MA).

According to the Brazilian Institute of Geography and Statistics (IBGE, 2010), the socio-economic development of Estreito can be divided into two distinct phases in time and space: (i) the first is related to the Tocantins River as an important means of social, economic and cultural articulation and/or integration between the communities of Estreito, which were territorialised along the river with other communities located in the north and northeast of Brazil; (ii) the second way in which the town of Estreito developed is related to the construction, in the second half of the 20th century, of the Belém-Brasília motorway, linking the states of Tocantins, Maranhão, Pará and others, enabling, at least for the towns located along this federal motorway, considerable and centralising economic growth.

Other large-scale projects built through the Growth Acceleration Programme (PAC 2) - which had an impact on the city of Estreito - were the implementation of the North-South railway, which cut through the city latitudinally, and, finally, the construction of the Estreito hydroelectric plant, which submerged part of the territories of the municipalities of Tocantins and Maranhão through its 650km lake[2] .

The Estreito hydroelectric power station began construction in 2007. One of its eight turbines was switched on in 2010, under President Lula. In 2012, under President Dilma Rousseff, the eighth and final turbine of the Estreito hydroelectric plant was inaugurated (MINISTÉRIO DO PLANEJAMENTO, 2010).

The option to build large infrastructure projects as one of the viable measures for Brazil's development process is, as this case study has shown, part of various development proposals undertaken, at least for the purposes of analysing more recent Brazilian history, by governments such as Getúlio Vargas (1930-1945; 1951-1954), Juscelino Kubitschek (1955-1960) and the military itself (1964-1985), when each of the rulers, in their own way, influenced by national and international elites, sought to develop Brazil through industrialisation.

Faced with clearer and more expressive economic and industrial growth from the 1970s onwards, and at the same time threatened with further development due to the global oil crisis, Brazil was forced to pluralise its energy matrix, opting for the construction of large hydroelectric projects. From this perspective, "the choice was made to base the country's industrial development on mega-electricity, taking advantage of the potential of the large number of existing rivers to build dams", as Benincá (2011, p.30) explains.

In making the case for the construction of large hydroelectric projects as one of the viable alternatives for maintaining the development of its economy, Brazil opted to deconstruct a social structure based on an agro-export economy in order to build a new Brazilian economic model, the industrial economy (CANO, 1985).

With industrialisation consolidated on a regional scale, as mentioned by Cano (1985, p.79), among others, the country demanded heavy investments from its rulers in the creation and expansion of basic infrastructure, because, for such a socio-economic context, "[...] the production of electricity, for example, could not meet the basic demand of Rio de Janeiro and São Paulo" (SKIDMORE, 1976, p.31).

As Brazil progresses with the construction of large hydroelectric power stations to support its new economic model, the first riverside communities affected by the consolidation of this electricity sector are also emerging regionally. The first social movements of riverside communities affected by dams emerged in the south, north and north-east of Brazil as a result of the construction of the first large-scale power stations. Since its inception, the Movement of People Affected by Dams (MAB), which was formed in the 1970s, has been made up of a heterogeneous social nucleus of riverside communities of fishermen, boatmen, farmers, indigenous people, in short, groups who, in one way or another, have been affected socially, economically, culturally, environmentally and territorially by the implementation of major hydraulic engineering works (BENINCÁ, 2011).

Even though they are different in terms of their formation and constitution, those affected by hydroelectric dams aspire to a single goal: to receive compensation that makes it possible for them to survive in other places where they are usually compulsorily relocated. According to Câmara (2013), one of the obstacles faced by those affected by dams in terms of compensation for material losses is the great difficulty encountered by both the companies responsible for building the projects and the public bodies responsible for licensing the construction of the plants in not knowing how to properly conceptualise who is in fact affected and deserving of probable compensation and who is not, even though their livelihoods have been disrupted as a result of the implementation of a hydroelectric project.

In view of this new socio-economic, political and structural situation that Brazil was going through, among the countries that were growing the most economically through their industrial process, the solution found in the medium and long term to continue keeping its economy solid and heated in energy terms was to prioritise the use of water from its river basins through the implementation of large hydroelectric plants, Historically, the first hydroelectric plants were built in the 19th century, as Benincá (2011) states, in the municipalities of Diamantina and Juiz de Fora, in Minas Gerais, with the aim of exploiting diamonds and providing public lighting, respectively.

The construction of these large hydraulic projects was accompanied by a long and cunning process of deterritorialisation of fishermen, boatmen, extractivists, barraqueiros, farmers, vazanteiros, among others, who, having formed riverside communities over several decades based on a close relationship with the great rivers, were gradually relocated compulsorily through compensation, in some cases to the most different places.

In this way, this model remains in line with the development policy pursued since the time of the military (1964-1985), which aimed, among other things, to analyse the technical, economic and social viability of hydraulic exploitation of the hydrographic basins[1] , mainly in the north of the country, in the process of expanding and strengthening Brazil's energy matrix.

Thus, as part of a policy of (diversified) structuring of the country's energy matrix, planned since the second half of the 20th century, the project to build the Estreito hydroelectric power station on the Tocantins River went from being a mere government aspiration to becoming, as of 2007, one of the largest hydroelectric projects sponsored by the federal government (REPÓRTER BRASIL, 2013).

In general, the construction of major infrastructure projects (building and paving interstate highways, railway lines, ports, airports and the implementation of dam projects) are all or mostly financed by the federal government itself, at a time when "the state began to operate as a financing agent for the construction of hydroelectric dams, supporting and subsidising large companies through the National Bank for Economic and Social Development (BNDES)", as Benincá (2011, p.31) points out.

As part of this developmentalist proposal by the Brazilian government, by forming a 650km lake[2] , the Estreito dam has socially and economically disrupted between 2,167 and 5,000 families, according to the Estreito Energia Consortium (CESTE) and the Movement of People Affected by Dams (MAB), who depended directly and indirectly on subsistence activities carried out on the Tocantins River. Although there is no agreement between the company building the dam and the Movement of People Affected by Dams on the number of riverside communities affected by the construction of this hydroelectric project, it is known that the failure to recognise some communities as unaffected also implies the non-payment of compensation for material damage caused by the implementation of this water project.

In view of the above, the problem guiding this research is based on the following assumption: To what extent may the social, political, economic and cultural assets of the riverside communities socially and economically impacted by the Estreito hydroelectric plant have influenced the compensation process?

For sociologist Bourdieu (2000), the position of an agent in a given society is determined and/or defined according to political, social, cultural and, above all, economic capital. It is precisely the complexity of these capitals that defines the privileged placement or distribution of individuals in society, since they determine the weak or strong ties that agents establish between themselves in order to get along in the context in which they are inserted, given that, in this capitalist logic, the impetus of the strongest prevails.

In order to achieve the objective outlined in this research, we prioritised: the application of structured and semi-structured questionnaires to the communities affected by the Estreito hydroelectric plant; the universe

---

[1] According to the Brazilian Institute of Geography and Statistics (IBGE), a hydrographic basin is a body of land bathed by a main river and its tributaries. In the case of the Tocantins-Araguaia basin, the Tocantins is the main river and the Araguaia is its main tributary.

of analysis in this investigation is made up of subjects belonging to five communities, namely fishermen, boatmen, farmers, stallholders and traders; in this way we sought to listen to both leaders and non-leaders affected by the hydroelectric project and those belonging to some of the communities mentioned above. Even though we didn't analyse indigenous populations, we recognise and mention them through some of the works by Lamontagne (2010), Sieben (2012) and Castro (2009), whose focus of analysis in homogenising terms can be characterised by the numerous socio-environmental impacts caused by the construction of the Estreito dam on indigenous populations.

As Duarte (2002, p.3) argues,

> [...] defining the criteria according to which the subjects who will make up the research universe will be selected is essential, as it directly affects the quality of the information from which it will be possible to build the analysis and arrive at a broader understanding of the problem outlined.

Social contact with the community through fieldwork is crucial when seeking to understand a given problem by listening to the individuals who experience it on a daily basis, because through their speeches, desires and experiences, one can more clearly understand the real dimensions, in the case of those affected by hydroelectric projects, of the socio-economic, cultural and environmental damage they have suffered as a result of the construction of these major engineering projects.

This research was preceded by the application of structured and semi-structured questionnaires, followed by the analysis of some documents, such as the Letter of Motion of Condolence provided by the leader of the Estreito (MA) and Aguiamópolis (TO) stallholders, showing the indignation of this community in relation to the social and economic damage that these individuals have suffered as a result of the construction of the dam. We also had access to some of the Public Civil Actions provided by the Tocantins Federal Public Prosecutor's Office regarding those affected by the Estreito hydroelectric power station.

As part of this methodology, we also used a field notebook[2] , which allowed us to obtain more information about the riverside communities impacted by the Estreito project. Based on this approach, we sought to gain a broader understanding of the types of conflicts the communities suffered and the criteria used by the company building the power plant, the Estreito Energia Consortium (CESTE), when it came to the compensation process.

In this way, it was possible to trace the profile of the riverside communities affected by the Estreito hydroelectric power station and, above all, to understand the socio-political process of the compensation paid by the consortium responsible for the construction of the hydroelectric power station in the middle Tocantins.

In building the theoretical-methodological basis for this research, we prioritised, among other authors:

---

[2] In this investigation, the notes in a field notebook served as a kind of extra diary. As the questionnaires were administered, individuals also made notes about the types and depth of the impacts they experienced as a result of the implementation of the Estreito hydroelectric plant.

Bourdieu's (2000) discussions about the social, political, cultural and mainly economic representations of agents in defining their positions relative to the context in which they are inserted; Granovetter's (2007), for whom agents make their socio-political decisions taking into account an entire social structure; as well as Abramovay's (2004) perspective, somewhat similar to those previously adopted by Bourdieu and Granovetter, of how social interactions govern markets and define them, also analysing it as something that is governed through the socially established interactions between the many different agents that make it up; Thompson (1998) and his analysis of the 18th century English market, stressing the strength of the socio-political movements of the hunger riots that characterised England at the time, making clear in his analysis the great political force of social mobilisation of these groups whenever deprivation or scarcity of some basic foodstuff knocked on their doors because of the unscrupulous action of some merchant, middleman, producer or even the inertia of the paternalistic state in relation to these issues.

As such, this dissertation has been organised as follows: In Chapter I, we present some discussions about the construction of hydroelectric power stations in Brazil, highlighting some issues related to the social, economic, environmental and cultural impacts that the large dams have had on riverside communities that live or used to live on the basis of subsistence activities carried out near the large rivers. In Chapter II, we address issues relating to regionalisation and the socio-political, economic and environmental effects felt by riverside communities on the Tocantins River as a result of the territorialisation of the Tucuruí, Lajeado and Estreito hydroelectric dams. We also alluded to the theoretical basis and the main theories that reinforce the importance of economic, political and cultural relations and representations between social actors in defining the relative position that each individual or group occupies in the social context in which they are immersed.

In Chapter III, we present the results and discussions about the socio-economic impacts and the process of socio-political construction of financial compensation for the communities affected by the Estreito power plant. Finally, in the final considerations, we summarise all the work carried out, including the conclusions and data that point to future research on the subject.

CHAPTER 1

SOME DISCUSSION OF THE SOCIO-ECONOMIC EFFECTS OF THE CONSTRUCTION OF HYDROELECTRIC PLANTS IN BRAZIL

Brazil, as a country of continental dimensions with an area of more than 8 million square kilometres and a large number of hydrographic basins, according to the Brazilian Institute of Geography and Statistics (Appendix B), has aroused the economic interest of governments since the second half of the 19th century with regard to the motivation for investment and reinvestment in the electricity sector with the use of large rivers to generate hydroelectricity for the most different Brazilian regions.

By prioritising the hydrographic use of its basins in the generation and distribution of electricity, and this more clearly from the 1970s onwards, when its industrialisation process demanded it more strongly, Brazil began to diversify its energy matrix in the name of progress and economic development, since in order to continue to maintain its level of development, it was necessary to discover new viable and sustainable energy sources from a socio-economic point of view.

The 1970s and 1980s in Brazil were characterised by the rush to carry out technical feasibility studies on a number of river basins, located basically in all Brazilian regions, and also by the implementation of major water projects, the ultimate aim of which was to maintain the pace of growth of the country's economy, given that, at the time of the boom, its position was one of the most privileged in South and Latin America, as Skidmore (1976, p. 31) and other authors point out:

> Since 1940, Brazil's GDP had been growing at 6 per cent a year, something that few Third World countries could match. Both Brazilians and foreign observers, noting the abundance of resources of almost every kind, predicted a bright future for Latin America's largest country. But further development would not be easy because the basic infrastructure was woefully inadequate.

To support its economic development model, Brazil began to prioritise the construction of large hydroelectric plants such as the binational Itaipu, installed in the south of the country through partnerships between the governments of Brazil and Paraguay in the 1970s and 1980s on the Paraná river, the dimensions of whose environmental, economic and, above all, social impacts have so far not been fully quantified, but are estimated at around 40,000 people.000 the number of people, according to Matiello (2005), who were compulsorily evicted as a result of the construction of this enterprise.

As a preventative measure to continue stimulating and favouring Brazil's economic and technological development, the government decided, in addition to building Itaipu, to invest in studies that could bring to light the possibility of exploiting other Brazilian hydrographic basins, prioritising those located in the North and Northeast regions due to the large number of flowing rivers found in these parts of the country, which had hitherto been overlooked for major investment in the energy sector.

Alluding to the great water potential of the Brazilian states grouped by federal regions, with regard to the exploitation of their river basins, Zitzke (2007) shows, as represented in the figure below, the percentage importance of each region in the Brazilian scenario, in which the North explicitly tends to overlap.

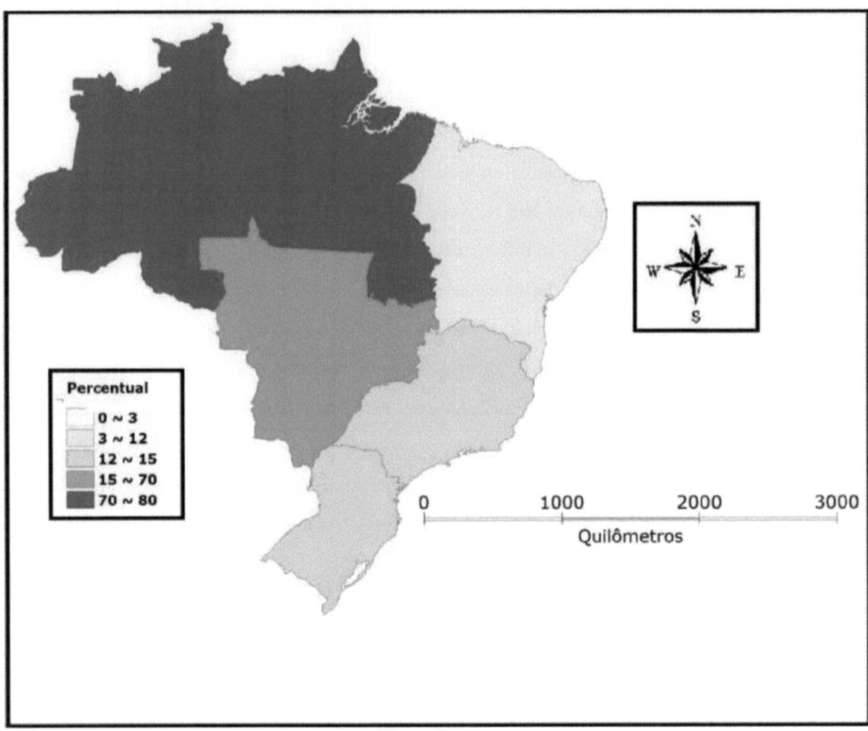

Figure 1- Percentage of water power in Brazil's regions.

Source: Zitzke. Adapted.

The implementation of these large dam projects, erected practically in every region of Brazil to provide economic and structural support for regional development, brought with it a series of political, social, economic and environmental problems for the riverside communities that depended directly on the great rivers for their survival.

> The construction of dams creates social and environmental paradoxes. In the name of development, countless families are affected and largely harmed, often not even being able to enjoy the benefits of the energy generated in the places from which they were expelled. It should be noted that the rivers, fauna and flora are not merely elements of the biome, but have deep connections with the culture and tradition of those affected. (BENINCÁ 2011, p. 53).

In the case of the construction of the Sobradinho hydroelectric power station on the São Francisco river in the north-east of Brazil, as Zitzke (2007) and Benincá (2011) state, the social, economic, environmental and cultural damage was responsible for the compulsory displacement of more than 60,000 people who depended on the São Francisco river for their livelihoods and, because of this work, had to reorient and reorganise their ways of surviving elsewhere.

In the face of the irreparable losses suffered by the communities of the São Francisco valley as a result of the territorialisation of the Sobradinho plant, there was a better socio-political articulation of those affected by the project, when compared to the riverside communities affected by the Itaipu hydroelectric plant in southern Brazil, because "unlike what had happened in the south of the country, here there was not a specific form of organisation of those affected, but an inter-union articulation to carry on the struggle". (BENINCÁ, 2011, p.84).

In this sense, it can be seen that the socio-political awareness of those affected by dams, in terms of claiming their material rights that have been jeopardised by the construction of hydroelectric projects, is more than a recognition on the part of the companies responsible for implementing such projects, it is the result of a social construction triggered by the communities themselves who, in a matter of months or years, have seen their socio-political, economic and cultural structures disrupted as a result of the construction of a major hydroelectric project.

For the National Movement of People Affected by Dams (MAB, 2012), the communities affected by the effects of the construction of hydroelectric dams are not against the implementation of the plants, but reject the way in which the companies responsible for their construction appropriate the territory and evict them, failing to recognise their material rights, which are most often drowned in the formation of the reservoir of large hydroelectric plants.

The logic of big business in the electricity sector, which seems to have stalled in time and space when it comes to appropriating territory for hydroelectric exploitation, is precisely that which consists of relocating the communities located in the areas subject to submergence by the filling of the reservoir to places that are increasingly distant from their origins, as if these groups of people affected by the implementation of hydroelectric power plants were inanimate, ahistorical and a-cultural objects and therefore had no ties to their former water territories. Thus, according to Silva and Silva (2011, p.2):

> They are, therefore, endeavours that aim to appropriate and reproduce space under the economic and exploitative logic of natural resources, disregarding the populations that live and have some material or immaterial link with the place that will suffer its action.

In order to make way for hydroelectric dam projects, big business needs to dismantle real social structures made up of riverside communities whose interests and perceptions of the riverbed, which will be interrupted immediately with the implementation of the plant, contradict the interests and conceptions about the use of this watercourse held by the so-called Joint Stock Companies, such as the Estreito Energia Consortium, for example, because the link with the Estreito region, which is now influenced by the hydroelectric plant, is consecrated to an exclusive type of socio-political exploitation of its water potential.

In this way, the transition from the subsistence territory of the riverside communities to that of big business takes place in conflict between the parties involved. These territories, which before the arrival of the project were basically controlled by these communities, formed a specific and almost endemic social structure of communities that survived on the multiple socio-economic activities carried out on the banks of the river.

From this perspective, according to Bourdieu (2005), every economic field or arena is made up of agents who conflict, dialogue and establish relationships, as they are part of a socio-political structure whose ability to do well in this game of interest will depend on the social ties enshrined in the economic, political and cultural capital that individuals possess.

When an energy company sets up in a particular geographical region, its objective has historically been to harness the water resources of that location, most of the time not caring about the negative social impacts that large-scale projects have had on riverside communities. From this perspective, Zitzke (2007, p.93) argues:

> Over time, the construction of reservoirs in Brazil has been based on isolated technical and economic decisions by a single sector, disregarding the other ways in which water is used at the project site, a fact that has led to many conflicts. The specific nature of a hydroelectric plant requires a multi-sectoral approach from its conception, planning, operation and maintenance, in view of its social, political, economic and environmental influence, considering its power to attract new investments to a given region.

From a developmentalist point of view, in order to territorialise a given region that is potentially suitable for hydroelectric exploitation, the companies responsible for building large-scale projects need to get rid of the social, political, economic, cultural and environmental structures of traditional communities, which until then had survived on their relationship with a given river, which, in this game of divergent interests, enters a phase of impact with the assembly of the project's infrastructure. When capital imposes itself on the water-geographical region with the aim of exploiting its natural wealth, it clashes with the multiple interests of the riverside communities that are fighting for their material rights, which are jeopardised by the implementation of the plant. Big business, through the capitalist state, seeks to impose itself at any cost on all individuals who come to be seen as obstacles to be removed if the project is to be territorialised. Thus, according to Sevá Filho (2008, p.5).

> The citizens affected, and the natural and built heritage that will be destroyed by the works, are seen as "interferences" in the studies and opinions guided by blind hydroelectric reason. The fact that there are people with possessions and rights working in the area, to be respected, and heritage to be defended, is stigmatised as an obstacle.

In the same vein, for the World Commission for Dam-Affected People (World Commission on Dams, 2000), the damaging effects that large hydroelectric projects have had on the lives of millions of people every year, and all over the world, are immeasurable and inhumane, if we consider that the companies responsible for the project exclude from the compensation process a large part of the riverside communities impacted by the construction of large-scale plants. For this Commission:

(A) Between 40 and 80 million people have been physically displaced by dams around the world;

(B) millions of people living downstream of dams-particularly those who depend on the natural functions of floodplains and fisheries-have also suffered serious damage to their livelihoods, and the future productivity of resources has been jeopardised;

(C) many displaced people have not been recognised (or registered), resettled or compensated;

(D) In cases where there was compensation, this was almost always inadequate; and in cases where the displaced people were duly registered, many were not included in the resettlement programmes;

(E) those who have been resettled have rarely had their livelihoods restored, as resettlement programmes generally focus on physical change to the exclusion of the economic and social recovery of the displaced;

(F) the greater the magnitude of the displacement, the less likely it is that the livelihoods of the affected populations can be restored;

(G) even in the 1990s, in many cases the impacts on downstream livelihoods were not adequately assessed or considered in the planning and design of large dams (WORLD BARRIERS COMMISSION, 2000, p. 20-21).

What is striking in this document produced and discussed by the World Commission on Dam-Affected People, with regard to the riverside communities affected by large hydroelectric dams, is precisely the unique and derisory type of financial compensation, when it happens, which is earmarked for a small group of people densely affected by the implementation of hydroelectric projects of great physical and spatial development.

Thus, the subsistence territory of the riverside communities, with their cultures, histories, beliefs and values, ceases to exist as a result of the territorialisation process of the medium and large hydroelectric enterprise, whose link with the region impacted by the shaping of the plant's lake is enshrined in a type of sole and exclusive exploitation of the riverbed that will be drastically modified with the implementation of the plant. For Zitzke (2007), this tendency to set up large hydroelectric plants, taking into account only the technical-economic aspect, which represents the capitalist sector of the energy industry, has been responsible for a series of conflicts in which HPPs have become territorialised. In analysing these social and economic conflicts generated by the construction of the Estreito plant in the communities mentioned above, Melo and Chaves (2012, p. 5-6) found that:

> The construction of the Estreito-MA HPP has generated countless discussions and, consequently, conflicts with the society affected, since there are countless people who live in the area surrounding it. This society has a close relationship with the river, mainly in terms of maintaining their diet based on fishing and, above all, in its ebb and flow where they produce the subsistence agriculture necessary to maintain their families.

Thus, considering the socio-spatial magnitude of the impact of the Estreito HPP, its territorialisation took place through the compulsory deterritorialisation of communities in both rural and urban areas that, in one way or another, had direct[3] and/or indirect[4] contact with the Tocantins River. For Chaves (2009), the rural and urban areas affected by the dam have similar values, as the following table shows.

Table 1 - Total number of people affected by the Estreito hydroelectric power station

Populations affected by the Estreito hydroelectric plant	
Urban	Countryside
1.148	1.019

Source: Chaves

After the Estreito Energia Consortium (CESTE) - made up of the companies Suez Energia Internacional, Vale do Rio Doce, Alcoa and Camargo Corrêa - won the tender to build the Estreito HPP, the next step in this developmentalist proposal was to "carry out the EIA/RIMA work, which began in 2001", as analysed in the Environmental Impact Report (2002, p.l). The region of Estreitense, compromised by the shape of the hydroelectric plant's lake, densely inhabited by indigenous communities, fishermen, boatmen, extractivists, farmers, ranchers, and others, almost all of them experienced psychological and social dramas as a result of the way in which the technicians responsible for the study imposed themselves on these individuals. Farmer Maria Natividade, who lives in the town of Estreito, emphasises this drama well:

> When it came to paying compensation for the damage we had suffered, the company's technicians said that the value of our property would be so much, and that was it, anyone who wanted to appeal could, but it wouldn't do any good, if we didn't accept the money they were paying us, they would come back here with the police and take us away by force[5]

In agreement with this farmer affected by the Estreito power plant, with regard to the way in which the expropriation process took place, the Federal Public Prosecutor's Office (Ministério Público Federal/TO) filed a Public Civil Action[6] against CESTE, claiming that it had imposed "real humiliation on the residents of Ilha de São José and PA Formosa, with regard to the dignity of the human person, forcing them to move before the settlement project was completed".

As the region affected by the Estreito hydroelectric project has led to the disruption of the livelihoods of more than 2. 000 families, according to the plant's Environmental Impact Report (2002), whose way of life was strictly related to the Tocantins River, this consequently led to a series of conflicts between the Estreito Energia Consortium (CESTE), the company building the plant, and the communities of fishermen, barraqueiros, barqueiros, who had a direct relationship with the river in question and who, since the first construction sites for the project were set up in 2007, have been fighting for some kind of financial compensation for the social and economic damage they have suffered.

---

[3] For the purposes of this research, we call direct impacts those that directly affect the communities that survive from primary and tertiary activities carried out on the bed or banks of the Tocantins River.

[4] Indirect impacts, for our purposes, are those that indirectly affect communities or groups of individuals such as the small traders at the municipal market in the town of Estreito who, before the plant's lake was created, basically bought their products from the vazanteiros, farmers, ranchers, fishermen, among others, who make up the group of people directly impacted by the Estreito hydroelectric plant.

[5] Oral report made in the field notebook on 22 November 2012.

[6] Document provided by the Federal Public Prosecutor's Office of Tocantins.

The developers of the Estreito hydroelectric power plant used a territorial-patrimonialist form of compensation (more details in chapter three of this research) when paying the communities affected by the plant and, as part of this process, compensation for socio-economic losses occurred in proportion to the type of impact, the link established with the affected site and the size of the impacts that the properties suffered as a result of the shape of the reservoir. Based on these three mechanisms: the type of impact, the link to the place and the size of the impact, the Estreito Energia Consortium defined the way in which its policy of compensating the impacted communities would take place.

In the specific case of this hydroelectric plant, compensation based on the territory that has been impacted can be defined, which is why it is called territorial-patrimonial, subdivided into four other types of compensation. According to data from CNEC Engenharia S.A for 2001 (CASTRO, 2009, p.71-74):

(a)   Assisted indemnification will be the alternative for non-owning families (squatters with or without permission) on properties with no viable remnants. This procedure consists of paying the families for the improvements they have installed on the property and offering them a support system for their reintegration into their homes and production elsewhere;

(b)   Rehabilitation of Remnants is the alternative that consists of avoiding the compulsory displacement of families who have partially drowned plots and who have remnants that are viable for them to remain productive or in conditions superior to those in force before the properties were affected;

(c)   Resettlement for Peri-urban Projects consists of setting up resettlement projects on the outskirts of urban centres close to the area of origin for families living in unviable properties who have work or employment relationships in the cities with agricultural activities carried out on the rural plots;

(d)   Resettlement for Rural Projects will be implemented in order to develop small-scale agricultural projects for the resettlement of impacted rural families.

The results of Castro's (2009) research into "the socio-environmental criteria for replacing losses and relocating those affected by dams", carried out on the community of Pahnatuba, a village in Babaçulândia, made clear the impacts suffered by this region, resulting in the process of dismantling the economic activities that the people of Pahnatuba carried out through direct contact with the Tocantins River or indirectly in its vicinity through lowland, extractive and subsistence agriculture. In his conclusion, Castro (2009, p. 19) recalls that "there is no price on everyday life, traditions and the loss of the lived environment".

Research such as that by Chaves (2009), Castro (2009), Gomes (2007), among others also mentioned throughout this investigation, is unanimous in confirming the negative effects that the option to build large hydroelectric plants has had on the countless riverside communities dependent on the relationships established with rivers with large volumes of water, such as the Tocantins, which since at least the second half of the last century has been undergoing major transformations as a result of the countless dams that the construction of hydroelectric plants requires in their implementation and operationalisation.

Ultimately, the effects of these large-scale projects on the lives of thousands of communities that are impacted practically every year in Brazil is just one facet of the countless injustices that the electricity sector has caused in the process of displacing riverside populations, whose survival is directly related to the many primary activities carried out on the banks of large rivers. As a result, the impacts they suffer as a result of these engineering projects are proportional to the type of activity each individual carries out and the link they have with the river, whether permanent or temporary, in the vicinity and/or on the receiving river itself.

What chapter three of this research makes clear, using the Estreito plant as a particular study, is that the territorialisation of large power plants doesn't just affect the physical-environmental framework of riverside communities, but instead ends up compromising an entire socially constructed structure in the vicinity of the Tocantins River, also interfering with the level of well-being of the communities directly affected by the construction of the Estreito hydroelectric plant.

In chapter two below, we seek to regionalise the socio-economic impacts that some large hydroelectric projects on the Tocantins River have tended to have on the riverside communities that depended on this watercourse for their survival through the countless economic activities they carried out on the banks of the Tocantins River.

CHAPTER 2

REGIONALISATION OF THE SOCIAL, ECONOMIC AND ENVIRONMENTAL EFFECTS OF HYDROELECTRIC DAMS ON THE TOCANTINS RIVER

The Tocantins-Araguaia hydrographic region has an area of 918,822 km^2; , which is equivalent to 11% of the national territory, covering the federal states of Tocantins, Pará, Goiás, Mato Grosso and the Federal District, as shown in figure 2. The configuration of this river basin is elongated in a north-south direction, following the predominant direction of the main watercourses. The Tocantins and Araguaia rivers unite in the northern part of the region, called the Tocantins River, which follows its course until it flows into the Bay of Marajó Island in the state of Pará.

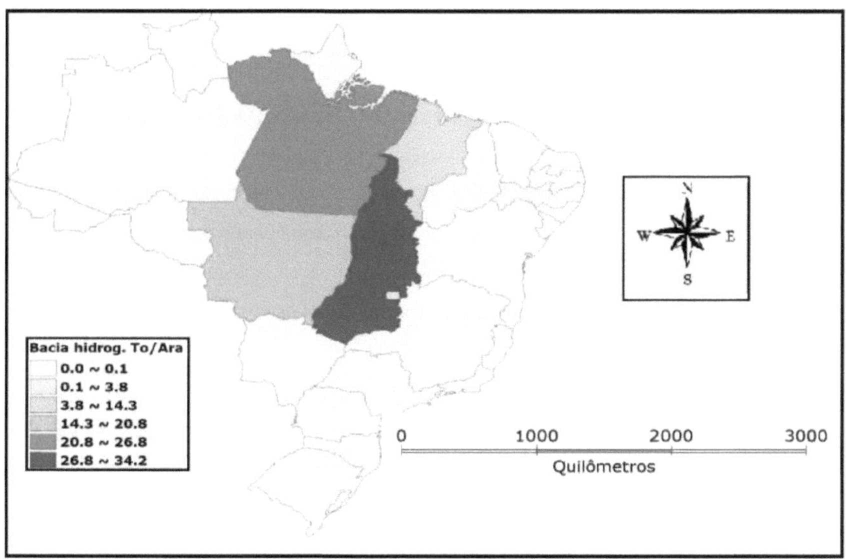

Figure 2 - Distribution of the TO/Ara catchment area in some Brazilian states.

Source: National Water Agency, 2013. Adapted.

This region, bathed by the Tocantins and Araguaia rivers, is home to the Amazon rainforest biome in the north and north-west, and the Cerrado biome in the other areas. Deforestation in the region intensified from the 1970s onwards, with the construction of the Belém-Brasília motorway, the Tucuruí hydroelectric dam and the expansion of farming and mining activities.

In 2010, around 8.6 million people lived in the Tocantins-Araguaia hydrographic region, which corresponds to 4.5 per cent of the national population, 76 per cent of them in urban areas. The demographic density was 9.3 inhab./km, much lower than the country's (22.4 inhab./km2) (AGÊNCIA NACIONAL DE ÁGUAS, 2013).

As Brazil progressed with the implementation of its first large hydroelectric power stations in the South, North and Southeast regions, the technical and economic feasibility study for the probable exploitation of the Tocantins-Araguaia hydrographic basin followed at all costs, studies which, in the view of Chaves (2009, p. 108):

The great availability of water resources in the Tocantins-Araguaia basin (in the north of the country) was emphasised, not forgetting that this type of developmental planning for the use of water in this basin began in the 1970s under the military governments, since the construction of the Estreito hydroelectric power station project has been rumoured since that time. To this end, Eletrobrás mapped out each basin in the Amazon region in order to determine the likely areas for the construction of hydroelectric projects.

With the theiritorialisation of these large hydroelectric projects, followed by the study for the construction of others in practically every region of the country, we can observe, in parallel to these events, the formation of a large mass made up of riverside communities gradually (de)territorialised from their ways of surviving along the great rivers, as a result of the implementation of these sectoral energy policies.

To give you an idea of the scale of the social, economic, cultural and environmental impacts produced by large power stations on the lives of thousands of families, the three largest hydroelectric power stations in operation in northern Brazil alone, Tucuruí, Lajeado and Estreito, caused almost incalculable damage to riverside dwellers, given that these were families who were generally well structured on the banks of the Tocantins River, the main river in the Tocantins-Araguaia basin, as shown in the table below.

Table 2 - number of people/families affected by the construction of the Tucuruí, Lajeado and Estreito hydroelectric power stations.

Hydro-energy projects	States	Number of families and/or people affected
Tucuruí	Pará	40,000 people
Lajeado	Tocantins	4,130 people
Strait	Tocantins and Maranhão	5,910 people

Source: Matiello and Castro. Adapted.

In this millennium, the consequences for communities directly impacted by the construction of large-scale sectoral energy policies tend to be incalculable from a social and economic point of view for those who suffer the negative effects produced as a result of the implementation of hydroelectricity generating projects. Analysing the effects of these projects on the lives of countless riverside communities, Vainer (1996, p.184) notes that:

Almost always carried out in peripheral regions, they have imposed rapid and profound changes on people's livelihoods and ways of life: compulsory displacement of thousands or tens of thousands of people, disruption of economic activities and labour and land markets, disruption of social relations, influxes of people putting pressure on already precarious infrastructure and service networks, changes in water quality, the course and regime of rivers with serious consequences for both health conditions and economic activities (fishing, lowland farming).

Following the implementation of the three largest hydroelectric projects in operation in the north of

Brazil (Chart 2), a large number of riverside communities classified as fishermen, farmers, boatmen, barqueiro, among others, experienced first-hand the hardships of the social, environmental and economic impacts on their livelihoods. Compulsorily relocated to the most diverse locations, with totally different characteristics when compared to the riverbanks, these riverside communities still deal with the great chance of not adapting to such environments.

Although we are not part of the riverside communities analysed in this investigation, considering that we would need much more time for a possible ethnographic survey of the impacts suffered, as has been done by some of the studies consulted, we cannot deny that the construction of the Estreito dam has produced profound changes in the intergroup, social, cultural, environmental and economic structure of the indigenous populations located in the region impacted by the shaping of the dam's lake. According to the national agency Repórter Brasil: "The four indigenous peoples on the banks of the river: Krahô and Apinajé, in the state of Tocantins, and Gavião and Krikati, in Maranhão, are against the dam and consider themselves impacted. None of them were included in the impact studies" (REPÓRTER BRASIL, 2013).

The effects produced locally by the construction of this project reflect the socio-environmental impacts generated on indigenous populations, according to Almeida's analysis (2007), and affect the different social actors involved in the implementation of the project in question, as they perceive the environmental licensing process for the hydroelectric plant. To carry out his studies, Almeida (2007) considered the conflicting visions and interactions of the environmental licensing process between six actors socially intertwined with the construction of the Estreito HPP:

- IBAMA, the Brazilian Institute for the Environment and Renewable Natural Resources, which acted by issuing licences for the construction and operation of the plant, considering that it is an undertaking that covers more than two states of the federation;

- FUNAI, the National Indian Foundation, was created with the aim of providing support to indigenous populations and also, as Almeida (2007, p. 26) argues, "to manage indigenous heritage, in the sense of its conservation, expansion and enhancement; to promote surveys, analyses, studies and scientific research on the Indian and indigenous social groups, among other objectives;

- CTI, the Centre for Indigenous Work, is a non-governmental organisation (NGO) set up to support indigenous groups. According to Almeida (2007, p. 33), "it is set up with the aim of reducing the dependence of indigenous communities on the state and other aid agencies, so that they take control of any and all interventions in their territories";

- CESTE, Consórcio Estreito Energia, an energy company set up by multinational companies to exploit the hydroelectric generating capacity of the Estreito hydroelectric power station on the Tocantins River in the state of Maranhão;

- MPF - Ministério Público Federal (Federal Public Prosecutor's Office) (2011, p. 2), whose work in relation to the implementation of a large hydroelectric plant, an undertaking that potentially impacts the environment, is centred, among other duties, on "[...] defending and protecting the environment, including all its aspects"; Finally, the sixth group of agents that Almeida (2007) emphasises in her study concerns the Apinajé, Gavião, Krahô and Krikati indigenous populations, because, for this researcher, these four indigenous groups, which together constitute one of the social actors impacted by the Estreito hydroelectric dam, became the focus of her dissertation investigation.

What is evident in Almeida's research (2007) is precisely the comprehensive, interactionist and politically visible analysis of the social agents impacted by the Estreito hydroelectric project, with their totally different conceptions of the use of the Tocantins River as a natural resource for multiple use, and not just for water use, as defended by CESTE and the state, the two politically best matched agents in the Estreito region.

Another study that mentions the impact of the environmental licensing process for the Estreito power plant, with a focus on socio-environmental conflict[7], interethnic[8] and multicultural[9], is the research carried out by Lamontagne (2010), in which the researcher sets out to understand how the state, as an agent promoting economic development, problematises and internalises the issue of the socio-environmental impact experienced by the social actors who experienced it as a result of the construction of the Estreito hydroelectric project.

For Lamontagne (2010), in the region affected by the implementation of the Estreito hydroelectric power station, the capitalist state, as the social agent responsible for the process of economic growth and/or development, has a dual purpose or action: firstly, it needs to stimulate the development of the economy through the implementation of major infrastructure projects (hydroelectric power station, duplication or construction of national integration highways, etc.), opening up a range of opportunities for the private sector to invest; and secondly, it needs to provide assistance to the populations that will suffer the negative impacts of the construction of such major public policies.

Sieben (2012) tries to understand the role of the state and the energy policy of the Estreito hydroelectric plant in the process of deterritorialisation of the community of Palmatuba, seen until then as a neighbourhood or extension of the town of Babaçulândia, located in the state of Tocantins. As an extension of the town of Babaçulândia, but with its own distinctive characteristics, Palmatuba sustained itself by extracting clay, babassu coconut *(prbignyamartiand),* small crops and livestock, fishing and handicrafts on the left bank of the Tocantins River.

In this research, the process of territorialisation of the Estreito hydroelectric project, one of the most polished phases of current capitalism in the energy sector, by consolidating itself in the far north of the state of Tocantins, generated, by extension, another process, the deterritorialisation of the community of Palmatuba,

---

[7] In this case, it's a question of conflicts over the perception of multiple interests in how to exploit a given natural resource, be it water, minerals, plants or whatever.

[8] To summarise, Lamontagne's research (2010, p. 26) refers to a type of interethnic friction. For further clarification, see the author's study.

[9] For Lamontagner (2010), this term arises between citizens of different cultural identities, often based on ethnicity, race, gender or religion.

which, like any other community located in the most diverse cities, was impacted by the construction of such a public energy policy. The communities have felt first-hand the many different types of negative impacts that a hydroelectric dam can generate. Thus, "the community of Palmatuba fell apart, which is why there is a need for a more in-depth study of the lives of these people, in order to understand how they adapted to the new situations they faced" (SIEBEN, 2012, p.16).

In disrupting thousands of riverside communities that depended directly on the Tocantins River for their survival, one thing has not escaped notice: it is not, when it happens, a question of mere compensation for purely economic damage, because a whole socio-political structure is at stake, resulting from a long and close relationship with the river, since, as we can see, the livelihood of the various riverside communities depends on all the economic, cultural, social and environmental activities that take place on its banks. Corroborating this view, Silva e Silva (2011, p. 2b) observes that:

> These are often people who have left their homes and who have a historical relationship with the territories they occupy, which are the stage for their cultural, social and labour manifestations, where the logic of the industrial-financial capital of the contemporary world does not always prevail.

There doesn't seem to be any understanding on the part of the companies building hydroelectric projects on the Tocantins River, because it's not simply a question of seeking financial compensation for all the damage that the implementation of large hydroelectric dams has caused in a wide variety of communities, since the compulsory displacement to which they have been subjected, however "just" it may be, has been responsible for the loss of cultural, social and, above all, economic identity of an ever-increasing number of entire families who are victims of the implementation of large-scale water projects.

When referring to the process of how companies build dams and appropriate a territory already inhabited by countless communities for the implementation of their projects, and how these communities have been politically and socially armed throughout the process of building large hydroelectric dams, in order to confront the companies themselves, if possible, in the fight for the recognition of their compensatory rights for all the damage that the plant can generate, Wadman (2002, p. 79) explains that,

> In all cases, authoritarianism and disrespect for the environment were present. The drama experienced by the affected populations, a wide range that includes indigenous nations, rural workers, river dwellers and various other traditional populations, found its political expression in social movements that began to directly confront the hydroelectric projects of the Brazilian state, expressing their rejection of a power generation policy that threatens the most legitimate interests of the populations involved.

With Brazil's political re-democratisation after 1980, riverside communities affected by the implementation of large power plants found the strength to fight against the very process of compulsory expropriation that characterised the 1970s, when the country was ruled by the military. Alluding to this new Brazilian phase and the socio-political strength that the riverside communities affected by the hydroelectric dams built up throughout this process, culminating in the democratic Brazil of the 1980s, Vainer (1996, p. 185) explains:

The First National Meeting of Workers Affected by Dams and the First National Congress of Workers Affected by Dams, held in Brasilia in April 1989 and May 1991 respectively, express the progress of these movements and the constitution of a political subject that began to intervene in an increasingly expressive way in the decision-making process and the implementation of policies in the electricity sector.

According to Foschiera (2010), the political awareness of the communities that suffer the direct impacts of the construction of hydroelectric dams, and this as a result of the socio-political construction of its members, through the Movement of People Affected by Dams (MAB), and the Pastoral Land Commission (CPT), through the progressive wing of the Catholic Church, were the main mechanisms responsible for the political formation of these new subjects in the arena of demands on the Brazilian scene, especially with the return of democracy in the country.

From North to South and East to West of Brazil, by forming socio-political movements as an alternative to exert pressure on the companies building the hydroelectric dams, the main objective of these affected communities has been to get these companies to recognise their social, economic and environmental rights, which are compromised by the execution of the hydroelectric dam and, based on these observations, to carry out compensation that includes all the affected communities, and not just those with land titles.

In the case of the Estreito hydroelectric dam, the focus of this research, from the time the Environmental Impact Study (EIS) and the Environmental Impact Report (RIMA, 2002) were drawn up, there were already a series of problems related to the areas subject to flooding and, consequently, possible compensation. In other words, at first, the quota stipulated by the study was 158m, which excluded some municipalities in Tocantins below this quota from possible compensation; later this quota was changed and maintained at 56m, which reduced the area that would be submerged by 10.8 per cent. As the study points out "[...] this was equivalent to a cut in generation capacity of approximately 70 kilometres, leaving the plant with an estimated capacity of 1,050 MW" (ENVIRONMENTAL IMPACT REPORT, 2002, p.9).

By constituting themselves as politically organised actors with common goals, the communities affected by the large hydroelectric plants began to consider and defend their environment, which is subject to the impacts of the plants, as part of their social, economic and cultural demands, as we will analyse in this next section.

2.1 The hydro-energy issue and socio-environmental problems in Brazil

Brazil in the 1960s and 1970s was characterised by a series of social movements that challenged the established order, movements that called for a more liberal, just and sustainable society, albeit in minorities and marginalised segments. These were, in fact, "movements with autonomous features, such as those of women, blacks, the environment" etc. (GONÇALVES, 2006, p.1 1).

With regard to the socio-environmental movement, in an attempt to provide a satisfactory response to the demands of the riverside communities, which were affected by the construction of the large dams and were beginning to realise the importance of their livelihoods, in this context, not only from an economic point of view, but, the Brazilian government, represented by the National Department of Water and Electricity, began to demand, for the construction of new HPPs, an Environmental Impact Study followed by an Environmental Impact Report (EIA/RIMA), respectively, as Scarlato and Pontin (1999) point out.

On the one hand, the Brazilian government was pressured by the ecological movement to review its development policy based essentially on the exhaustive exploitation of forest, mineral and water resources and, on the other hand, the government was instructed to comply with some of the determinations of international organisations with regard to the environment, if it wished to continue to be favoured by the policies for acquiring foreign loans, as Vainer (1996) explains when referring to a series of social movements with a contestatory basis, especially the ecological movement,

> At the same time as it was facing the emergence of these movements, the electricity sector was under pressure from both ecological movements and multinational financial agencies (IDB, World Bank), which began to impose environmental requirements in order to grant credits (VAINER, 1996, p.1 85b).

In the midst of this socio-political situation, which was shaken in contestatory terms by various social movements, some political parties emerged, such as the Workers' Party (PT) and especially the Green Party (PV), as a kind of response and at the same time political and organisational support for marginalised groups and those socially and environmentally excluded from the development policy adopted until then, as mentioned by Benincá (201 1) and Ridenti (1992), among others.

Obviously, the protest movements that emerged in Brazil in the 1970s only found solid foundations to consolidate themselves in the 1980s, when the democratic regime of government was re-established, followed by the promulgation of the 1988 Federal Constitution. This Constitution, unlike the previous ones, in its article 225, establishes clear and precise limits on the role of public authorities and the community in matters concerning the care of the environment, which is now treated as the common good of all, who are responsible for defending and preserving it for present and future generations.

Backed by the 1988 Federal Constitution, social movements, Non-Governmental Organisations (NGOs), public and private entities and the mass media have all found full legal support since the promulgation of the Constitution to fight for the protection of the environment in all its aspects, including riverside communities that are socially and spatially affected by the implementation of large hydroelectric plants. In this same line of argument, "social movements now have a new guise, going beyond labour relations, which were previously targeted, to diverse issues such as defence of the environment, among others", as Melo and Chaves

(2012, p. 3-4) attest.

The regionalisation of the large hydroelectric projects in the Tocantins-Araguaia basin brought with it a series of socio-economic impacts for the riverside communities that had lived on the banks of the Tocantins River for decades. In order to expropriate these communities, which were included in the quota subject to submergence, the Estreito Energia Consortium used the policy of compensation based on the title of ownership of the land that would be totally or partially affected by the formation of the plant's lake, which, at the other extreme, ended up exempting other communities that did not own land from this same process. From this perspective, according to Melo and Chaves (2012, p.l):

> The territorial impacts of the construction of the Estreito HPP reservoir are causing countless changes, affecting the territorial identity of the local community, especially in terms of its relationship with the Tocantins River, both economically and culturally.

Based on these assumptions of the collective impacts generated for the communities affected by the Estreito hydroelectric power station, the theoretical and methodological foundations of this work seek to understand the mechanisms used by each community investigated in this research during the process of compensating their improvements affected by the Estreito power station reservoir, based on one question: How might the social, political, cultural and, above all, economic assets, referred to by Bourdieu (2000) as capitals, have exerted some kind of influence on the financial compensation of the communities affected by the power plant project? This and other questions are what we sought to understand in the following topic.

22 **Theoretical and methodological bases**

From Bourdieu's perspective (2000), it is essentially the combination, depth and breadth of a series of capitals (cultural, political, symbolic, economic and social) that will determine the relative position of individuals in the society in which they are inserted. As beings resulting from the most different cultures, historical by extension, human beings relate in time and space through an endless set of social networks that are subject to the most different socio-political arenas in a game of interest, whose decision-making measures, which weigh up the advantages that each community will enjoy, is exercised by the breadth and depth of its capital in the territory of its materialisation. Bourdieu (2000, p.134) thus defines "agents and groups of agents [...] by their relative positions in this space".

Thus, applying this same principle to the field or social arena of the riverside communities affected by the Estreito HPP, we will try to understand the extent to which the economic, political, cultural and social capitals mentioned by the French sociologist, in their interactionist complexity, have influenced and/or are influencing the compensation process of the many different riverside communities who, in one way or another, have suffered some kind of loss or gain as a result of the construction of the Estreito hydroelectric power station in Maranhão.

In the representation of the communities affected by large hydroelectric projects, even though they are dealing with a single space (physical/territorial), the perception of this space in cultural, economic, social and environmental terms, by each affected community, does not represent the same thing (meaning) as if the way of survival of the people affected by these projects were the same.

ranchers, barraqueiros, fishermen, boatmen, farmers and traders, who make up the groups analysed in this research, were a homogeneous and linear whole organised according to the same socio-economic objectives and interests.

Each social representation of the communities affected by the implementation of the Estreito HPP is organised according to the socially constructed ties and the weight (verticality and horizontality)[10] that each of the economic, political, cultural, etc. capitals pointed out by Bourdieu (2000) represents in the existential maintenance of each group in the regional context where they are located. For the purposes of this investigation, it is important to know how this set of capitals may have been decisive in the process of establishing and claiming possible financial compensation for material damage caused by the implementation of the Estreito HPP.

In fact, we are talking about a socio-political dispute represented by CESTE, as big capital, and the communities affected by the plant, representing a type of diffuse and fragmented capital. In this respect, we have a "[...] struggle that is both theoretical and practical for the power to preserve or transform the social world by preserving or transforming the categories of perception of that world" (BOURDIEU, 2000, p.142).

In this way, a given geographical region is characterised by a series of capitals which, constituting socially established structures that are re-established by the strength of the agents that make it up, end up defining the social advantages that each individual enjoys in the environment in which they are immersed. Thus, by economic capital, we mean the way in which capitalist agents organise themselves in given social structures, through ties and relationships of conflict and reciprocity, trying to perpetuate themselves in power, whether through legal or illegal means, the success of which will depend on the articulation of economic capital with political and cultural capital (BOURDIEU, 2000).

In this logic of argument, symbolic or cultural capital would be "[...] merit earned, honour built, prestige, fame; likewise, political capital is understood as a field of forces and as a field of struggles aimed at transforming the relationship of forces that gives this field its structure at a given moment" (BOURDIEU, 2000, p. 134-135;165-166).

For some of the main exponents of economic sociology, such as Granovetter (2007), Abramovay (2004) and Bourdieu (2000), the space socially created and recreated by human action, the ultimate cause of

---

[10] For the purposes of this research, verticality means the social ties established between individuals belonging to a certain social class, such as fishermen, stallholders, indigenous people, among others, who make up one or several groups respectively; verticality, also for our purposes, means the legal and juridical ties established between groups and subgroups with the institutions and organisations that can represent them.

all types of materialisation of the most varied types of capital, economic, cultural, symbolic and social, is the result of a long process whose most striking mark is its condensation into the force (power) that agents have in the course of human activities.

Like Bourdieu (2000), Granovetter (2007) sought to uncover decades and/or centuries of scientific production in which the economic dimension had been analysed as something atomised, devoid of interaction with the other fields, political, cultural and social, which, for Bourdieu (2000), express the power that economic agents exert or represent in the social environment. For Granovetter (2007, p. 9):

> Actors don't behave or make decisions like atoms outside a social context, nor do they slavishly adopt a script written for them by the intersection of social categories they may occupy. Instead, their attempts to carry out purposeful actions are immersed in concrete and continuous systems of social relations.

Thus, for Granovetter (2007), the structure of the social environment, from the simplest to the most complex, cannot be defined solely by the economic dimension devoid of intersocial contact. In this same conception of analysing the economic field as an inseparable part of a structure that is socially articulated with all the other fields, Abramovay (2004), like Bourdieu (2000) and Granovetter (2007), seeks to deconstruct a conception of classical economics, for which the occurrence of economic activities is independent of a network of other socially established contacts between culture, politics, universities, and so on.

The economic order of any market, regardless of the relative time that history presents us with, has always resulted from interactionist actions established between the agents of the economy, culture, politics, entertainment in public squares, in short, it is the set of these social elements, and not just an extra-social economy, that defines the structures of human arrangements. However:

> It is true that contemporary economics lives up to its widespread reputation as a grey, mechanical science incapable of incorporating ethical precepts into its assumptions. However, an important and increasingly significant part of the discipline turns precisely to the study of concrete forms of social interaction and calls into question the purely *selfish* and maximising motivations postulated by the neoclassical tradition (ABRAMOVAY, 2004, p.l).

Regardless of the context and social structure, for the sociology that began to be produced from the 1960s and 1980s onwards, the market is one, and only one, of the countless socially defined dimensions in any communal structure. The new theoretical assumptions of Economic Sociology, which began to be prioritised mainly in the last quarter of the decade mentioned above, concern the way in which social scientists began to see the market, not as something separate from society, self-regulating from time to time according to historical periods and the forces of some autonomous economic agents who, regardless of culture, politics and the countless social forces that act in divergence and convergence in a given social field or arena, were able to maintain themselves in a given social structure (BOURDIEU, 2000).

Along these lines,

Thompson (1998), when analysing the riots in 18th century England, clearly saw that the people who constantly rose up against the wheat growers, millers, bakers or even the paternalistic state itself, had a keen perception of the interdependent relationship between the economy, culture and politics. In Thompson's analysis of the social representations of the English economic model, it was observed that it had,

an ideal existence and, equally, a fragmented real existence. In years of good harvests and moderate prices, the authorities were forgotten. But if prices rose and the poor became turbulent, the model was resurrected, at least to produce the symbolic effect (THOMPSON, 1998, p.160).

For 19th century England, whose remnants of a paternalistic state were becoming an increasingly distant and difficult reality to maintain, social representations, which in Thompson's research were practically condensed into mobilisation, uprisings and popular clashes, were no longer justified with such vehemence in that century. While in the middle of the 18th century English riots were organised politically and socially to fight for their common goals, in the following century social rights were extended and, along with them, the political consciousness of the British, now represented by their trade unions, put pressure on industrial capitalists to fulfil their social duties. What is clear from Thompson's (1998) historical-structural approach is precisely the political and social strength that social movements gained over several decades of English economic and technological development.

Political awareness and the organisational capacity of social mobilisations are the result of permanent construction in the most adverse situations that societal life can bequeath. This political awareness that the English masses had built, and which condensed with all its vehemence in the 19th century (characterising an England that was no longer agrarian and/or agropastoral), now belonged to a country whose economic-industrial emergence had formed a thinking mass with a high capacity, through trade unions, to assert its social, political and ethical rights, which had historically been structured through the most varied social representations. As Thompson (1998, p. 186) states: "the hunger riot did not require a high degree of organisation. It required a consensus of support in the community and an inherited pattern of action with its own objectives and limits."

When it comes to the popular demonstrations undertaken by riverside communities affected by hydroelectric projects, even though they refer to another context and another type of claim, when it comes to fighting for the defence of their economic, social and environmental rights, those impacted by the current capitalist system have been able to mobilise, albeit in communities segmented by fishermen, boatmen, farmers, to build and enforce their socio-economic rights affected by the territorialisation of Brazil's large hydroelectric projects.

Since the construction of Itaipu on the Paraná River, Tucuruí, Lajeado and Estreito on the Tocantins

River, and lastly Belo Monte on the Xingu River, which is currently being implemented, it can be seen in the conceptions of Benincá (2011), Waldman (2002), among others, that there has been a kind of politicisation of the leaders representing the riverside communities affected by the hydroelectric plants, This politicisation tends to be passed on to the members of each community as the main weapon for mobilising and demanding their social and, above all, economic rights, which have been disrupted by the implementation of large-scale HPPs such as those mentioned above.

As we can see, human beings are part of real socio-political structures, characterised by advances and setbacks in relation to social conquests, as Bourdieu (2000) rightly pointed out when he discussed the importance of social, political, cultural and economic capitals as driving forces behind the privileges that individuals enjoy. It is these capitals, in the arena where they are represented, that will define the relative position of the most different individuals.

In the case of those affected by the Estreito hydroelectric plant, the area outlined for this case study, the social achievements that some riverside communities are enjoying, although reasonable, have materialised more clearly as a result of their socio-political articulations condensed into associations that represent them, such as the Association of Boatmen of Maranhão (ABEMA), Associação dos Barraqueiros da Praia do Pé da Ponte e Ilha do Cabral, Colónia dos Pescadores and the Movimento dos Atingidos por Barragens (Movement of People Affected by Dams), which, through numerous political actions, are trying to sensitise the riverside communities to fight for their rights to compensation for the socio-economic damage they have suffered as a result of the hydroelectric plant.

When he compares eighteenth-century England with the century after, Thompson (1998) discovers that the way in which the masses of people made their demands had changed and adapted to the socio-economic transformations that nineteenth-century England presented to the workers. Faced with this new context and politically aware of the importance of organising themselves through trade unions, because reality had now changed, workers knew perfectly well where to turn when they wanted to implement a social policy.

According to Foschiera (2010), when the Movement of People Affected by Dams (MAB) emerged in the 1970s, its basic aim was to bring together and represent the most varied riverside communities affected by the implementation of large-scale hydroelectric projects. In this way, those affected became politicised, which, according to Bourdieu (2000), becomes one of the capitals, political capital, represented, in the case of those affected, by associations, fishing colonies and the MAB itself.

In the following chapter, the procedures and results of this field research, referring to the social and economic impacts that the communities made up of fishermen, boatmen, farmers and traders had as a result of the construction of the Estreito HPP, are presented together with the respective compensation for material losses resulting from the territorialisation of this hydroelectric project.

CHAPTER 3

SOCIO-ECONOMIC IMPACTS AND THE SOCIO-POLITICAL CONSTRUCTION OF COMPENSATION FOR COMMUNITIES AFFECTED BY THE ESTREITO HYDROELECTRIC DAM IN MARANHÃO

3.1 The choice of study area

The study area, the city of Estreito in Maranhão, where the Estreito HPP is located, was chosen because of its importance in the country's current energy policy and, conversely, because this municipality is home to a large number of riverside communities classified as barraqueiros, barqueiros, fishermen, farmers, vazanteiros, indigenous people, among others, who were directly affected by the construction of the hydroelectric plant, and also because of the ease of access to this region. The representation of these communities in the research, and not others such as the Apinajé, Gavião, Krahô and Krikati indigenous populations, who were also affected by the dam, is justified by the fact that in November 2012, when the questionnaires were administered, we only found these communities in the city of Estreito, the location chosen for this study.

When the field research was carried out, 30 questionnaires were answered by six communities, whose percentage representation in the total sample was as follows: 27% farmer, 27% trader, 20% fisherman, 13% boatman, 7% barqueiro, 3% vazanteiro and another 3% who did not define themselves. The purpose of the questionnaires was to outline, in a more concrete way: the situation of the problem related to the social and economic impacts that these communities have been facing; and the proposals for financial compensation that the Estreito Energia Consortium (CESTE) proposed before and after the reservoir was created.

The questionnaires applied to the region's residents were characterised as structured and semi-structured, prioritising both those who received and those who did not receive compensation for material losses resulting from the construction of the plant in question. The questionnaires applied to those affected by the Estreito HPP aimed to verify:
- The criteria used by the company responsible for building the plant when compensating the communities;
- the mobilisation of communities with regard to their social and economic rights, which have been compromised by the plant;
- the degree of satisfaction of those affected with the compensation received;

- the degree to which the communities analysed depended on the Tocantins River for their subsistence activities;
- the level of improvement experienced by families in terms of well-being;
- social and economic changes due to the construction of the HPP;

By tabulating the data and cross-referencing the information, it was possible to associate the relationship between income and the compensation received, as well as the type of activity and educational level.

3.2 Geographical characterisation of the city of Estreito

Located in the south of the state of Maranhão, between parallels 06°33'38 South and 47°27'04 Southwest, the city of Estreito has a population of 35,738 inhabitants, according to the latest IBGE demographic census (2010). According to the IBGE, the historical process that gave rise to this city began with a small settlement that sprang up near the Tocantins River, whose political and administrative influence fell under the jurisdiction of the city of Carolina, also in the state of Maranhão. Given its long history of close ties with the Tocantins River, on 27 December 1954, the town of Estreito ceased, albeit temporarily, to be a district of Carolina, backed by State Law No. 1304, and was renamed Presidente Vargas, in honour of the then President of the Republic Getúlio Vargas. Almost three years later, more precisely in May 1957, the town of Presidente Vargas had its emancipation process cancelled by the Federal Supreme Court, returning it once again to the status of a district of Carolina.

Thus, in March 1982, still according to IBGE data (2010), the day before Brazil's political re-democratisation, through State Law 4.416, the town of Estreito regained its political and administrative autonomy from Carolina, returning to the status of a town and receiving the name Estreito.

With the development policy of then president Juscelino Kubitschek, the current town of Estreito, at the time a district of Carolina, ended up benefiting from this government policy, because with the inauguration of the Presidente Juscelino Kubitschek de Oliveira bridge on 2 January 1961 over the Tocantins River, linking the town of Aguiamópolis in Tocantins to the town of Estreito in Maranhão, social and economic relations between these two states and the others located in the north and northeast of the country became easier and faster. After the inauguration of this bridge and, in less than two decades, the Belém-Brasília motorway (1974), the town of Estreito experienced a fundamental and decisive economic upsurge in relation to the other towns and cities located in the south and north of Maranhão, and far from this national integration motorway.

33 Geographical context of the Estreito hydroelectric plant in the mid-Tocantins region

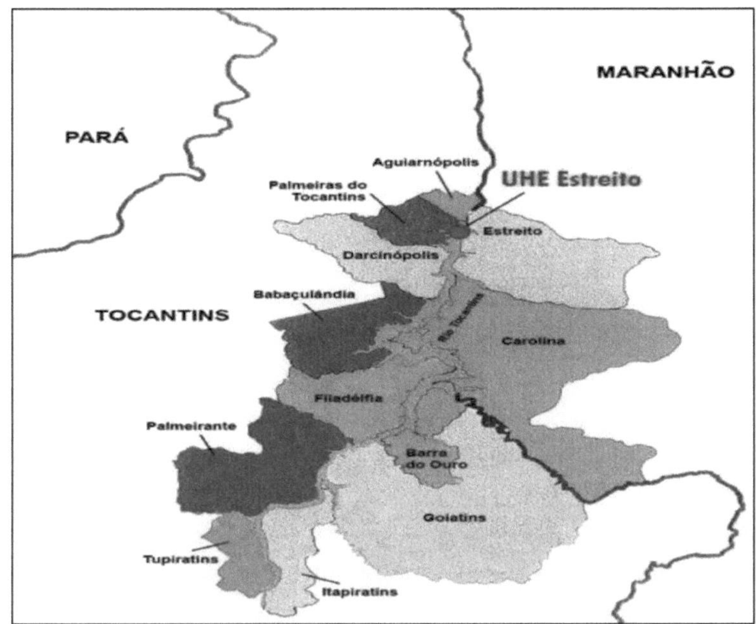

Figure 3 - Characterising and locating the Estreito study area in Maranhão

Source: Consórcio Estreito Energia (CESTE).

Planned since the 1970s, when the first technical and economic feasibility studies were carried out for its probable construction, the Estreito hydroelectric power station, located on the middle stretch of the Tocantins River between the municipalities of Aguiamópolis in the state of Tocantins and Estreito in the state of Maranhão, has, according to the news agency Repórter Brasil (2007), followed the following chronological order for its installation and operationalisation, as shown in Table 2.

Chart 3 - Installation and operationalisation schedule for the Estreito power plant

ACTION	MONTH	YEAR
The Estreito Energia Consortium wins the ANEEL auction.	July	2002
Obtaining the Preliminary Licence.	April	2005
Obtaining the Installation Licence.	December	2006
Construction work begins on the plant.	April	2007
Inauguration of the first turbine.	November	2010
Inauguration of the eighth and final disturbance Estreito hydroelectric plant.	October	2012

Source: Repórter Brasil (2007) and Ministry of Planning (2010). Adapted.

As this is a major undertaking involving the territory of the federal units of Tocantins and Maranhão, the reservoir of the Estreito hydroelectric plant, which covers a surface area of 650km, with a total generating capacity of 1109.70 MW, directly and indirectly impacts" the cities of Estreito and Carolina in the state of Maranhão, and in the state of Tocantins the cities of Aguiamópolis, Babaçulândia, Barra do Ouro, Darcinópolis, Filadélfia, Goiatins, Itapiratins, Palmeirante, Palmeiras do Tocantins and Tupiratins" (ENVIRONMENTAL IMPACT REPORT, 2002, p. 5-7).5-7).

As you can see, the shape of the plant's lake, covering part of the territory of two Brazilian states, clearly demonstrates the proportions of the social and economic impacts that the region of Estreitense, in the south of Maranhão, and Bico do Papagaio, in the north of Tocantins, suffered as a result of the implementation of this hydroelectric project. When analysing the impact that a large artificial lake, formed from the physical territorialisation of a hydroelectric plant, tends to generate locally and regionally, Branco (2004, p.1 12) argues that "the impact caused by a hydroelectric project derives essentially from the extent of the area flooded by the reservoir to be formed by the intersection of the river with the dam".

3.4 Discussion of results

Even though this is a continuous territory in physical terms, and multicultural in social and economic terms, the company building the Estreito power plant used two basic criteria in the process of compensating the affected populations, which even

While expressing different meanings, they converge on the same denominator, that is, compensation based on the land title held by the ranchers, farmers and ranchers and, at the other extreme, excluding from this compensation process the fishermen, barraqueiros and partially the boatmen, who for decades had maintained a close relationship with the Tocantins River.

It should be noted in this study that the names barraqueiro, barqueiro, fisherman, farmer, vazanteiro and comerciantes are social constructions defined by those affected by the Estreito hydroelectric power station when primary data was produced. Historically, they are terms created by social movements from the 1960s and 1970s onwards, when Brazilian society politicised left-wing movements that sought improvements and changes to the country's then developmentalist policies. As we have discussed throughout this research, the apex of social movements with a contestatory basis in Brazil, for the purposes of analysis in this research, coincides with the period when the country was under the aegis of the military and enjoyed galloping industrialisation, albeit on a regional scale.

For the purposes proposed, we consider, in terms of scale of scope, that social impacts include all other impacts, i.e. economic, cultural, territorial and environmental impacts that those affected by hydroelectric projects are subject to when a large hydroelectric plant is built. Generally speaking, those impacted by such sectoral hydroelectric policies are categorised into two basic types: those concentrated upstream (above where the plant was built) and those concentrated downstream of the project, i.e. below the reservoir. In both cases, we are talking about socio-economically heterogeneous communities that have been seriously impacted by the construction of large hydroelectric projects.

When we analysed the territory of the riverside communities in the city of Estreito in November 2012, the period in which the questionnaires that guided the development of this research were applied, we came across six groups of individuals classified as barraqueiro, barqueiro, fisherman, farmer, vazanteiro and small traders who were impacted by the Estreito HPP.

The barraqueiras communities, a contingent of approximately 30 families who carried out their economic activities downstream of the project, which are represented in figure 5 by 7% of the sample, were largely damaged by the implementation of the Estreito hydroelectric plant. Even though they were communities that carried out their activities only during the river's dry season, which was concentrated between the months of June, July and August each year, the material damage was responsible for the disruption of their economies practised directly on the beaches of the Tocantins River. As the stallholder Francisco Ramos da Silva Ribeiro Gonçalves argues:

> Before the dam was built, we worked for three months of the year on the beaches of Pé da Ponte between Aguiamópolis do Tocantins and Estreito in Maranhão, and also on Ilha do Cabral, right in the middle of the Tocantins River, when it was dry. Before the dam, my profit was between R$5,000.00 and R$6,000.00 a month.000.00 to R$ 6,000.00 per month, after the reservoir was filled it dropped to around R$ 2,000.00 to R$ 3,000.00 per month when people were on the beach more often[11].

In similar situations of socio-economic impacts generated by the implementation of the Estreito hydroelectric power station, but with a deeper scope in terms of social and economic disintegration, the case of the boatmen includes a contingent of approximately 48 individuals who, as shown in figure 6, represent 10% of the sample surveyed and who, unlike the other communities analysed in this study, were the ones who showed the greatest stage of economic disintegration, as shown later in figure 8, when we analyse the communities together. This community, whose relationship with the Tocantins River used to occur every month of the year, with the formation of the power station reservoir, which (abruptly) interrupted the traffic of people and goods on the river, ended up making it impossible to continue this activity, as these individuals can no longer navigate the Tocantins River as they did before the power station was built.

The boatmen practically have their boats in the dry because they can no longer travel freely on the Tocantins River, since the construction of the Estreito dam has put an end to decades of socio-economic development, the limits of which were determined solely by the river. As boatman Vicente de Paula Pereira de Araújo says:

> My profession as a boatman was passed down from father to son, my father taught me how to navigate the river, now, after they built the dam, my boat only stays dry all the time because in July we still take some bathers to the beach, my income was R$2,000.00 reais before the dam, now I only work basically one month a year with my boat, my income today is R$9,00.00 reais during the beach season. In order to survive, I do odd jobs here and there where work comes up[12]

In the specific case of this community, of the 48 boatmen who were regularly registered with the Maranhão Boatmen's Association (ABEMA), only 8 did not receive compensation, precisely those who were not affiliated and up to date with the association. Among the communities analysed in this research, by the time the field survey was carried out in November 2012, only the vast majority of boatmen and all the farmers had received compensation for the socio-economic impacts of the Estreito hydroelectric plant.

What varied among the farmers impacted by the dam in terms of the compensation process was precisely the amount paid to each family that had suffered socio-economic damage as a result of the project, as can be seen in figure 6, where the percentage of this community in the total sample is equivalent to 30 per cent. In this specific case, the company responsible for setting up the plant compensated each family based on the land title they held to their respective properties affected by the hydroelectric dam's lake.

With the socio-economic impact of their subsistence activities, the fishermen affiliated to the Estreito colony, a contingent of 190, who were regularly registered and up to date with their obligations, which consisted, among other things, of paying 5% tax to the colony on all the fish they managed to catch, paying an annual fee of R$10.00 (ten reais) for internal bureaucratic services and paying some of the colony's own employees. In the case of this fishing community, they had not received any compensation from the company by the time of the fieldwork in November 2012. The 17 per cent of these fishermen shown in figure 6 is equivalent to 30 individuals in the sample that guided this work.

At the moment, the only thing that CESTE has done to alleviate some of the suffering and social and economic inconvenience experienced by this community, in practical terms, is to keep a minivan with a driver at the disposal of the fishermen who carry out their fishing activities upstream of the plant. According to fisherman Raimundo Nonato Falcão Lima, who has been fishing in this region for 46 years, the company promised compensation of R$10,000.00 for each fisherman's family and a villa at the top of the dam, but so far it has not fulfilled any of its promises and, because of this, they are experiencing certain financial difficulties which, in some cases, have caused some fishermen to move away from the town of Estreito. Others have changed activities, working in various informal jobs, weeding plots and clearing fields on other people's farms, etc.

Another community that was socially and economically impacted by the shape of the dam's reservoir, and which also did not receive compensation for the material damage caused by the hydroelectric dam, was the vazanteiros, who carried out economic activities on the banks of the Tocantins River, but the land did not belong to them. This group represents 3% of the individuals in the six communities analysed in this research.

The small traders at the municipal market in the city of Estreito, who account for 30 per cent (figure 6), are part of the communities indirectly affected by the plant, as they don't carry out any kind of activity directly on the Tocantins River, but instead buy almost all the products they sell at their stalls from the vazanteiros, chacareiros and fishermen who ultimately make up the groups of communities directly impacted by the Estreito hydroelectric plant.

In addition to this approach, in an effort to gain a better understanding of the real social and economic impacts of the Estreito hydroelectric plant and, as a result, possible compensation, we carried out a joint analysis of all the communities affected by the plant. They can be characterised into two large groups of individuals who suffered the direct and indirect impacts of the plant, and can also be subdivided into smaller groups with the boundaries between them measured by the link that each community had with the river before the plant. According to the Environmental Impact Report (2002), these matrix groups are: a) landowners, b) non-landowners.

The first group includes landowners with land titles, classified as farmers, ranchers, farmers, and also part of the urban population that has suffered some kind of direct impact as a result of the shape of the power plant's lake, since this dam has totally and/or partially submerged part of their plots in the urban perimeter of the city of Estreito, which, by extension, includes them "in the Areas of Direct Influence - AID, also known as the socio-economic environment" (ENVIRONMENTAL IMPACT REPORT, 2002 p.23).

The non-owners included in the Indirect Areas of Influence (III) of the power plant lake, also known as the physical-biotic environment, can be classified as fishermen, barqueiro, boatman, tenant, vazanteiro, extractivist, etc. These are communities that, even though they didn't have land titles for the properties on which they worked permanently or temporarily, maintained some kind of subsistence link with the Tocantins River.

In the light of everything that has been presented, we have previously interpreted the data that was collected in the field, seeking to understand the dimensions of the socio-economic impacts on the daily lives of the communities that were affected by the implementation of the Estreito hydroelectric power station.

3.5 Communities and socio-economic impacts

The survey consisted of a total of 30 individuals, belonging to communities made up of fishermen, vazanteiro, barqueiro and traders, who were directly affected by the construction of the Estreito HPP. According to gender, they are distributed as follows: 6.67 per cent female and 93.33 per cent male, and the average age of those affected is 51 years. With regard to length of residence, those affected have lived in the same place for approximately 42 years. In terms of schooling, the sample had an average of 7.9 years of study, with a minimum of 0 and a maximum of 17 years of study, and a relative average lower than the number of years of high school study. The sample's level of education is highly variable, as the standard deviation of 4.06 characterises it in this way, but it was observed that the majority of those impacted by the plant have a level of education concentrated in incomplete primary education, and only a few of these have higher education and postgraduate degrees.

With regard to the family budget, those affected are heads of household and had a family income of R$1,682. This income was defined based on all the income that families earned over the course of a month, before the plant was built, from fishing, planting, transporting people or goods, or even selling a product on the streets or beaches, as is the case with stallholders and street vendors. In this specific case, it is important to emphasise that the family income generally comes from permanent subsistence activities that the head of the family carries out near or related to the Tocantins River. The presence of *outliers* is notorious in this variable, given that the value of the standard deviation is R$1,524.01, so there was a wide dispersion of values depending on the type of activity that each impacted person carried out and/or carried out on or near the river.

It's also important to emphasise that surveys with a small sample universe like this one, considering that it involves a contingent of 2,167 families in the Estreito Energia Consortium's count and/or 5,000 families in the accounts of the Movement of People Affected by Dams in the Estreito region, have a strong tendency to show a significant disparity in relation to the income variable.

The number of people per family was concentrated around 4 to 5 members, corroborating the figures presented by the Environmental Impact Report (2002, p. 63-64), which showed an average of 4.7 individuals for rural properties and 4.4 for nucleated families in the urban area, with a dispersion, as only one family in the sample had 11 members.

Table 1- Socio-economic characterisation of those impacted

Variables	Average	Standard Deviation	Minimum value	Maximum value
Age	51,1	10,65	31	75
Length of Residence	42,67	16,43	15	75
Education	7,9	4,06	0	17
Family Income	1.644,25	1.304,51	400	5.000
People per family	4,37	2,01	1	11

Source: Survey results, 2012.

In general, the basic variables, as shown in table 1, capture the behaviour of individuals in relation to their level of social well-being before the Estreito HPP was implemented. To determine this level, we adopted family income as the main parameter (positive or negative) of the quality of life of the communities impacted by the hydroelectric plant. In this case, it was noted that the level of well-being of families tended to decrease considerably after the construction of the project.

As for the level of satisfaction of the communities with the compensation they received, 35% of those affected were satisfied, precisely those who make up the group that received compensation: the farmers and almost all the boat people. On the other hand, 65% of those analysed in our sample were dissatisfied, which is equivalent to the representation of the other communities made up of fishermen, barraqueiro, vazanteiro and traders who did not receive compensation from CESTE for the social and economic losses they suffered. This fact is also justified by the type of activity the communities carried out before and after the construction of the HPP and, above all, whether or not the individuals owned the property where they worked. The previous premise was supported by the information collected, given that specific groups such as farmers and boatmen comprise the only communities analysed in this research that have received compensation to date.

To determine the communities' level of satisfaction with the implementation of the hydroelectric plant, the groups analysed in this study were asked about the social and economic advantages and disadvantages generated by the construction of the plant. Around 45 per cent of those impacted, representing those who received financial compensation, were satisfied with the hydroelectric plant, while 55 per cent of those impacted by the plant were dissatisfied with its construction, corresponding to those communities who were not compensated by the company.

Notes taken in the field showed that the advantages and disadvantages mentioned corresponded to the particular interests of each group, without a holistic view of the advantages generated for the whole town of Estreito. Satisfaction and/or dissatisfaction with the Estreito project was also justified by the subsistence practice that each individual carried out before and after the construction of the HPP, given that, afterwards, various activities carried out by some individuals were affected, which means that the vast majority experienced some kind of change.

Table 2 - Composition of the sample in terms of satisfaction with the compensation received as a result of the construction of the HPP.

	Yes (%)	No (%)
Satisfaction with the compensation received	35%	65%
Satisfaction with the construction of the HPP	45%	55%

Source: Survey results, 2012.

As a complement to the previous analysis, and in an attempt to understand the socio-economic impacts produced in the communities studied in this research, we also sought to analyse the importance of the Tocantins River in the development of the countless subsistence activities carried out by the communities on the banks of the river, as well as the types of relationship they established with it.

Analysing figure 4, it is possible to see how dependent families were on the Tocantins River to carry out their different types of economic activities before the dam was built, and in many cases the populations were concentrated around primary activities. Around 63% of the population surveyed considered their subsistence activities to be important, with permanent dependence, regardless of the type of activity, whether primary or not.

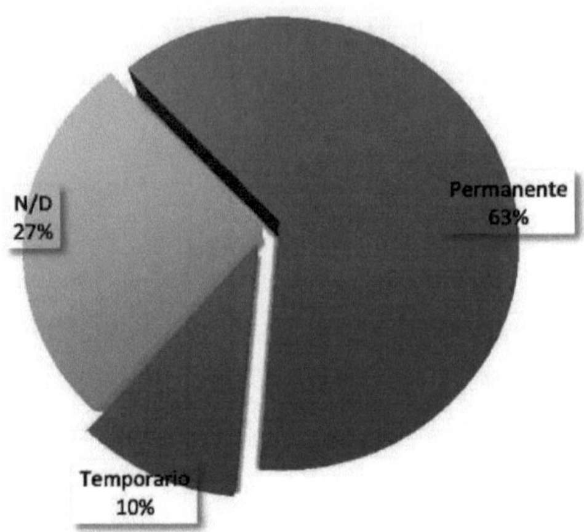

Figure 4 - dependence on the Tocantins River for economic activities carried out by communities located near the HPP.

Source: Field research, 2012.

The development of primary activities by the members of the families that make up the communities analysed in this study can also be described as follows

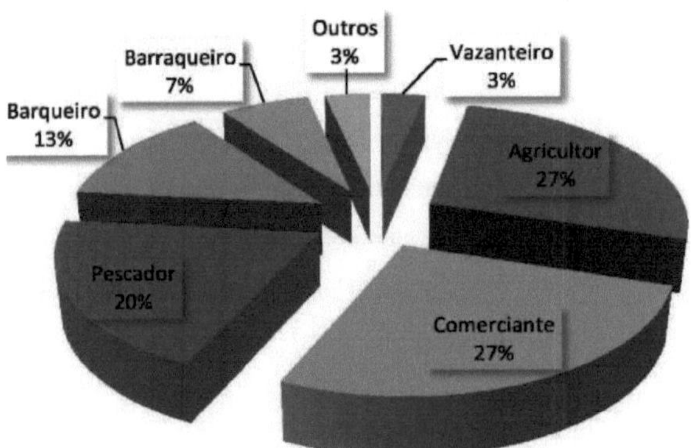

Figure 5 - Main subsistence activities carried out by affected families before the construction of the power plant.

Source: Field research, 2012.

We observed that commercial activities accounted for 27 per cent, followed by agriculture, also with 27 per cent, and fishing with 20 per cent of the entire sample region of this study. The percentage of these three activities gives us an idea of each individual's level of satisfaction with the construction of the HPP and, consequently, the level of satisfaction with the compensation received, as shown in Table 2.

It's worth noting in this context that the individuals who worked as fishermen and stallholders, who represent

33% of those impacted, did not receive financial compensation for their losses as a result of the construction of Estreito, and are therefore dissatisfied with the implementation of this hydroelectric project. Subsistence activities[13] were affected by the construction of the hydroelectric power station in a disorganised way, as they depended directly on the Tocantins River, which determined the right times to plant and harvest through its flood and flow regimes. Tertiary activities such as commerce, which in this case is carried out at the greengrocer's and at the municipal market, were only specifically and positively affected.

After the construction of the Estreito HPP, some changes were observed with regard to the activities carried out by the individuals living in this region. The subsistence activities carried out by each of the families after the construction of the Estreito HPP are illustrated below

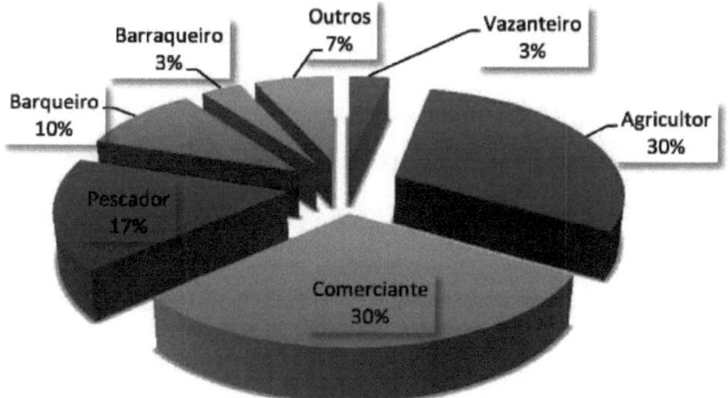

Figure 6 - Main subsistence activities carried out by the families affected after the construction of the power plant.

Source: Field research, 2012.

The most significant changes are seen in the activities carried out by fishermen, which fell from 20 per cent before the dam to 17 per cent after the project; there were also negative changes after the construction of the dam in the activities carried out by boatmen and stallholders, falling from 13 per cent and 7 per cent before the dam to 10 per cent and 3 per cent after the project, respectively. Considering that our sample consisted of 30 individuals in a universe made up of 5,000 thousand and/or 2,167 families, as expressed by MAB and CESTE respectively, these small percentage differences represent a sharp drop in their income levels and consequently in their quality of life, which was evident in the field research carried out in November 2012.

Analysing the sample representation of the fishermen and stallholders in the group of 30 individuals impacted in this research, the economic loss they have suffered as a result of the construction of the plant has forced them to partially change their occupation, i.e. some are currently working as shop assistants, bricklayers' helpers, motorbike taxis and weeding plots. This means that there has been a kind of socio-economic destabilisation of these families as a result of the construction of the Estreito hydroelectric plant.

---

[13] The term in focus refers to the activities that communities carry out on/in the banks of the Tocantins River, in other words, it can be synonymous with survival. Therefore, every time it appears in this research, it will be referring to this.

In line with our approach, Benincá (2011, p. 44) reveals that: "The effects of dams on people's lives are almost never properly measured and expected. They shape dramatic pictures of socio-environmental injustice". However, with regard to commerce and agriculture, it was observed that these activities experienced positive changes and remained in first place in importance, considering the types of activities practised by the communities investigated.

The level of satisfaction with the construction of the HPP can be partially explained. We observed positive or negative changes that some socio-economic cultures experienced after the construction of the HPP, given that the level of family income decreased, increased or remained stable in some cases. Figures 5 and 6 show that primary activities such as agriculture and fishing accounted for approximately 47 per cent of our total sample, which consisted of 30 individuals spread across five communities in the city of Estreito, Maranhão.

Figure 7 shows that these activities are carried out by a population with a low level of education, generally individuals with incomplete primary education. Trade is also concentrated in this region, practised by individuals with a low level of education, also with incomplete primary education. In addition, the theory associates the primary subsistence activities carried out by families in a given region with their low level of education, which can be seen in the figure below.

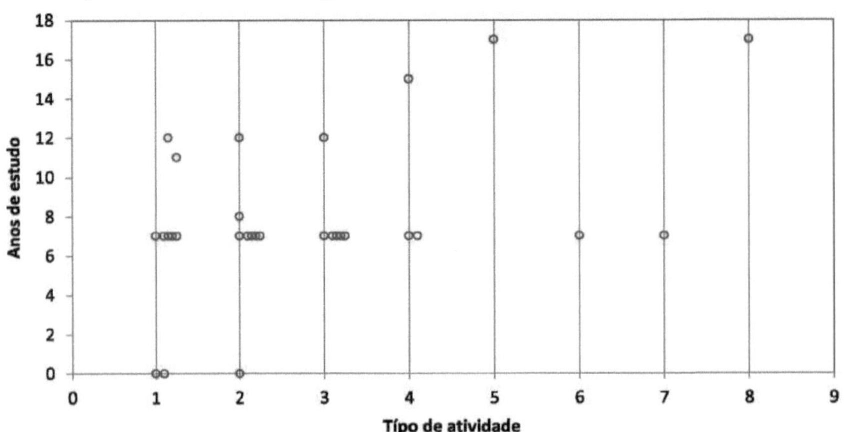

Figure 7 - Relationship between educational level and type of activity of those impacted
Source: Field research, 2012.
Classification of activities:
Activity 1 = Farmer; Activity 2 = Trader; Activity 3 = Fisherman; Activity 4 = Boatman;
Activity 5 = Barraqueiro; Activity 6 = Fazendeiro; Activity 7 = Vazanteiro; Activity 8 = Outros.

An overview of the consequences of the construction of the Estreito HPP on the well-being of families (figure 9) shows a succinct summary of the real social and economic damage that the main communities that depended on the Tocantins River have suffered as a result of the development. In detail, it can be seen that every activity that families carry out in this region has lost some importance, given that the level of well-being measured in this case by income has fallen, with the exception of traders and farmers.

40

The loss of importance in income is different for each community analysed, with percentage variations as low as 8% and as high as 49%, as in the case of the boatmen, which is why the level of dissatisfaction with the construction of the HPP is justified in previous comments. In the specific case of the boatmen, which justifies a separate caveat, and which does not exclude the possibility of further research to explain in more detail the huge financial losses incurred as a result of the construction of the dam, the dissatisfaction stems from the extinction of their weekly waterway routes after the construction of the reservoir, as shown in the figure below.

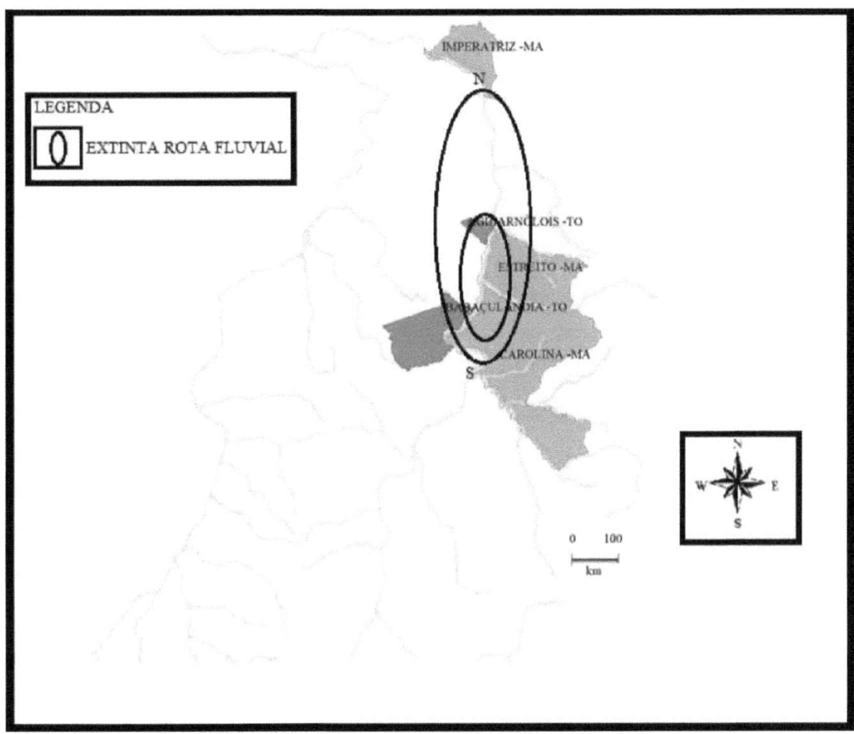

Figure 8 - Extinction of the old river-weekly waterway by the Tocantins River.
Source: Own elaboration based on field notebook notes, 2012.

The riverside communities living in the cities of Tocantins, Aguiamópolis and Babaçulândia, and in Maranhão, Imperatriz, Carolina and Estreito, as represented in this illustration, were drastically impacted commercially and financially by the establishment of the Estreito hydroelectric power station. This trade route, which enabled social, economic and cultural exchange between the residents of the above-mentioned cities, among others, was completely wiped out after the creation of the dam's reservoir.

In the words of Brandão, a boatman and resident of Ilha do Cabral, located between the towns of Aguiamópolis, in Tocantins, and Estreito, in Maranhão, one of the pioneers in navigating the Tocantins River via the waterway described above, and with more than 50 years of experience in the area.

41

years transporting people and goods from the island of São José[14] to the towns of Carolina and Babaçulândia upstream from where the Estreito HPP was built, the ebb and flow along these river routes between the states of Tocantins and Maranhão took place practically every day of the week, with a more pronounced concentration on Wednesdays, when the boatmen travelled upriver towards the island of São José and, on Fridays, travelled downriver, picking up people and goods on the banks of the Tocantins River to take them to the port of embarkation and disembarkation in the city of Estreito.

Also according to Brandão, the 1950s, 1960s, 1970s and 1980s were characterised at regional level by the consolidation of the river route: São José Island, Carolina, Babaçulândia, Estreito, Aguiamópolis and, finally, Imperatriz. Considering the complexity of the goods and people that routinely travelled along this waterway, connecting traders, farmers, vazanteiros, fishermen and others, the dissolution of this river route also diluted or de-structured decades or a century, in the case of São José Island, of survival on the banks or in the middle of the Tocantins River before the construction of the Estreito hydro-energy project.

According to the Estreito Environmental Impact Report (2002, p. 79): "Having lost the linear accessibility of the waterway, the connection between the riverside towns today is fragile and even non-existent." According to Chaves (2009, p. 114), "[...] this impact on the space in which these communities interact brings to light a series of conflicts that end up culminating in a process of struggle for their own survival and consequently for their existence as a traditional community".

Rivers, from the simplest to the most complex, for single or multiple use, have taken on a meaning of their own for the many different communities that have come to depend directly or indirectly on the natural benefits that these rivers can provide them. Their visions of what these rivers mean to them differ qualitatively and quantitatively from the concept of their use and importance held by the big Brazilian and international energy entrepreneurs. According to Benincá (2011, p. 27):

> While companies target the territory (the river) with economic interests, the resident communities see it as a space for sustainability and survival. They maintain a deep identification with the place, adopting it as an environment in which to live and socialise.

The importance of the Tocantins River as a means of integrating the North and Northeast regions of the country and its most central part, Goiás, dates back to around the 18th century, when gold mines were discovered. According to Lira (2011, p. 125): "The Tocantins waterway sustained the survival of the region, through trade, albeit seasonal, but maintained between the cities of the upper Tocantins and the states of Maranhão and Pará, through the ports of Imperatriz (MA) and Belém (PA)."

Navigation on the Tocantins River has made possible the social, economic and cultural development of countless riverside communities, which have established themselves sustainably along its banks. Thus, "The Tocantins River has been the main commercial freight transport route in the region over the last few centuries" (GOMES, 2007, P.19).

---

[14] The island of São José does not appear in the illustration as one of the centres where people and goods spread to the cities of Carolina, Estreito and Imperatriz because, according to Brandão, the Estreito hydroelectric power station completely submerged the island, which until then belonged to the city of Babaçulândia, Tocantins.

With the implementation of the Estreito hydroelectric project, between the states of Tocantins and Maranhão, thousands of families suffered the impacts of its construction. According to Chaves (2009), there is no consensus between the Estreito Energia Consortium (CESTE), the company responsible for building the plant, which puts the number of people affected at 2,167 families, and the Movement of People Affected by Dams (MAB), which puts the number at 5,000 families whose livelihoods have been compromised by the implementation of the Estreito HPP.

Other crops such as fishing, commerce and agriculture are significant in this context. It's important to mention that the percentage values shown in figure 8 below correspond to the averages obtained according to the real values of family income in each of the communities, and that there are variations and different situations in each family that makes up a group. For example, in the area of commerce, the analysis required an effort to understand the positive consequences at two different times: (i) before the plant was built; (ii) after the first turbine was inaugurated, when the project no longer required so many workers. Based on this assumption, there was a certain decline in the purchasing power of some traders, across the board, although there were some families who experienced improvements. This can also be seen in the practice of other activities, but in general the positive variations are not decisive.

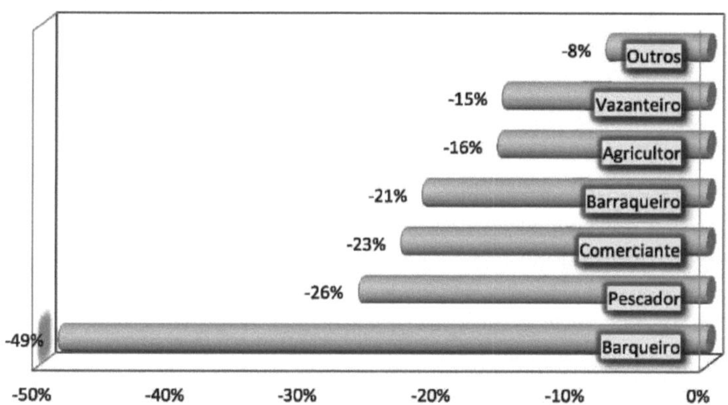

Figura 9   - Percentage change in family income after the construction of the Estreito HPP, according to the type of subsistence activity.

Source: Field research, 2012.

Figure 10 shows a relationship between the level of education and the income of individuals, showing that in some cases there is no explicit relationship between a high level of education and a high level of income. This fact was expected in this research, but was not confirmed, because in poorly developed regions, such as the one where the Estreito hydroelectric power station was built, in the south of the state of Maranhão, on the border with the far north of the state of Tocantins, a high level of education was not an advantageous assumption for proposing the hypothesis of a high level of income.

It can be seen that the highest concentration of income is around R$1,000, a level associated with people with more schooling (incomplete primary education), but in turn there are people with the same income or higher, in some cases, but with zero education (no study or schooling), corroborating the premise mentioned above.

On the other hand, there are individuals with a high level of education and a high level of income, which specifically shows that the level of education is related to income, but that in this case study the relevance of this group is small and dispersed.

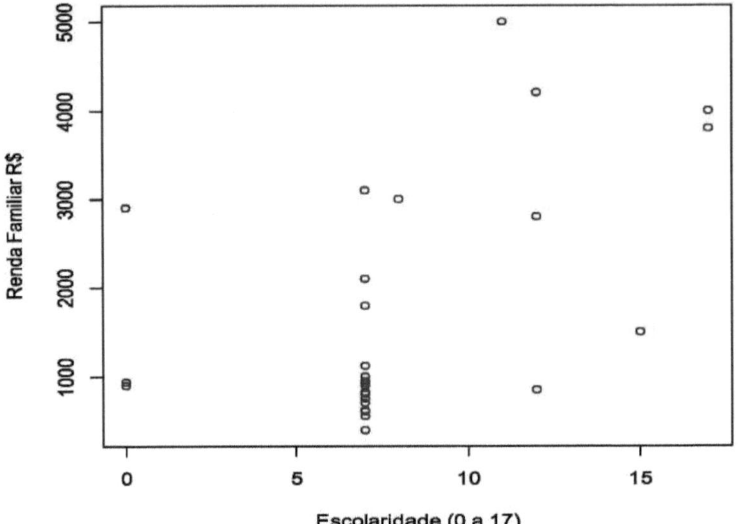

Figura 10 — **Relationship between family income and level of education of those affected.**
Source: Field research, 2012.

3.8 The communities affected by the Estreito hydroelectric dam and the compensation process

The Movement of People Affected by Dams (MAB) emerged in the 1970s as a direct response to the development policy of the Brazilian government, which in the current political and economic climate had opted for the construction of large hydroelectric dams, barqueiros, vazanteiros, farmers, squatters, among other social groups who were physically, socially and economically affected by the implementation of the first large-scale hydroelectric projects such as the Itaipu and Tucuruí dams, both built in the late 1970s and mid-1980s, as Chaves (2009) and Foschiera (2010), among others, point out.

One of the great challenges to be overcome in contemporary society when it comes to communities affected by hydroelectric projects concerns the lack of a clear and objective definition of how many and which riverine social groups are actually affected by hydroelectric dams and subject to probable financial compensation for material losses resulting from the formation of reservoirs. Thus, according to Câmara (2013, p. 3-4):

44

The discussion at the moment between the state, the companies responsible for the works and the MAB is centred on divergent points regarding the conceptualisation of who these people are, how many there are and the amount of compensation to be paid to each community that has been identified in the socio-territorial spaces subject to submergence due to the implementation of the plant. In the case of the Estreito hydroelectric plant, with regard to the compensation process, it was observed that the criterion adopted by the company that won the bid, the Estreito Energia Consortium (CESTE), was the territorial-patrimonialist bias, a policy historically adopted by companies that build large hydroelectric plants in which the state's contribution in this process is enshrined in a type of abstention, giving construction companies total freedom to negotiate directly with the communities affected by the hydroelectric plant.

In the fight for recognition of their compensation for material damage, those affected by the Estreito Hydroelectric Power Plant held a large march of approximately 100 kilometres between July and August 2010, from the city of Araguaína, in the state of Tocantins, to the city of Estreito, in the state of Maranhão, where they had set up camp in front of the hydroelectric power plant for over a year.

And so it is for Santos, an vazanteira and fisherwoman for more than 25 years in the town of Estreito, and a member of the march to recognise compensation for the communities affected by the hydroelectric dam:

> We were about 400 river dwellers when we left the city of Araguaína for our lodgings in front of the power station, marching between five o'clock and eleven o'clock in the morning because of the intense heat. By the time we arrived in front of the dam, the movement had grown stronger and there were already around two thousand people, including fishermen, farmers, river dwellers, barqueiro, boatmen and farmers, all affected by the dam and who had not yet received any compensation from the dam[15].

According to Benincá (2011), the compensation received by communities whose livelihoods are included in the flood quota can basically be classified into three different types:

(a) landowner compensation, in which only the owners of the flooded land are considered to be affected;

(b) water compensation, includes only the entire population displaced by the dam;

(c) physical-economic indemnity, in which all communities, landowners and non-landowners alike, who have had their economic activities jeopardised by the construction of hydroelectric power stations, are considered to deserve compensation for material losses.

---

[15] Oral report made in the field notebook on 13 November 2012.

As Silva and Silva (2011, p. 3) argue:

First and foremost, those affected are those who somehow feel the effects of the construction and operation of a hydroelectric power plant, and may or may not be compulsorily displaced to areas other than those they used to occupy. Or they may be displaced to areas that are geographically distant from where they used to live.

When we analysed the compensation processes for the communities affected by the Estreito HPP in the Middle Tocantins, we observed that the company responsible for carrying out the project, the Estreito Energia Consortium, followed the logic of patrimonialist territorial compensation, which took into account two basic principles when it came to financial compensation: (a) the size of the properties and (b) the improvements inventoried by the company's technical staff, as shown in the table below.

Chart 4 - Description according to the size of the properties of those affected by the Estreito HPP and their respective compensation.

OWNERS	HECTARES	INDEMNIFICATION
Francisco José de Araújo	13	R$ 142.000
Humberto Santana de Brito	65	R$ 75.000
Pedro Vanderlei de Souza	7	R$ 45.770
Cleomar Vanderley Chaves	4.5	R$ 45.000
Antônio Raimundo R. da Silva	10	R$ 42.000
José Dias de Souza	6	R$ 30.000

Source: Field research, 2012.

The fact that a farmer with less land received more financial compensation than another who had a larger property is justified by the fact that these properties had improvements; another point that complements this discussion is the fact that, in some cases observed, a small producer had one or two children with their own homes living on his property. In the latter case, only the father-owner received compensation in cash, leaving his children with another type of compensation, i.e. land of between 6 and 10 hectares, but no longer on the banks of the Tocantins River, and a Vale Compra Card worth between R$120.00 and R$180.00, valid for one year.

Also part of the communities compensated by the Estreito Energia Consortium were around 98% of the boatmen directly affiliated to the Estreito do Maranhão Boatmen's Association (ABEMA), founded in 1995. On the other hand, fishermen, barraqueiros and vazanteiros were left out of this process, i.e. communities that, even though they had a direct link to the Tocantins River and suffered directly from the impacts produced by the plant, on either the Maranhão or Tocantins side, were left out of the compensation process because they were not considered owners or had title to any land.

In addition to the economic, cultural, social and environmental damage that the implementation of the Estreito power plant has caused to the countless riverside communities analysed in this research, another aggravating factor that was observed during the fieldwork concerns a total of ten families who were fishermen and vazanteiros at the same time, who were not compensated by CESTE because they didn't own their vazantes, but who, after the dam's lake was formed, their homes were in areas of constant risk. In other words, with the damming of the river, their humble homes built of adobe and babassu coconut straw (Annex G) were only about 200 metres from the edge of the lake, and about a kilometre downstream from where the dam was built.

For the vazanteira and fisherwoman Bezerra:

> Before the dam, our houses were about 700 metres from the bank of the Tocantins river, after the plant the distance dropped to about 200 metres, we are subject to flooding if one of the hydroelectric dam's floodgates breaks, even in these conditions, the dam technicians responsible for compensation say that we will not receive financial compensation because the place where we live belongs to the federal government (bank of the Tocantins river), and the place where we cultivated our fields belongs to the farmers who have already received compensation[16].

The degree of impact and consequently the compensation received by the communities affected by the construction of the Estreito Hydroelectric Power Plant was determined according to the owner-dependent relationship that the communities of farmers, fishermen, barraqueiro, vazanteiro, barqueiro and others had with the Tocantins River. What the interpretations of the field data revealed about the socio-political process of material compensation for these communities was that, in general, more than 60 per cent of them had an extremely interdependent relationship with the river, deriving their main livelihood from it and its flooding/leakage periods, and after the reservoir was built they were unable to do the same.

In a letter addressed to the Institutional Relations Manager of the Estreito Energia Consortium, Isac Braz Cunha, Maria de Fátima Silva expresses her indignation and that of the other stallholders affected by the plant:

> We are humble human beings, but we are not fools and we will fight for our rights until the last day of our lives. We realise that we won't be able to carry out the beach seasons as we have done for over 20 years. We won't work with mobile tents, so that every day when the river fills up, we can go up with it. Nor will the population swim in a river with violent waters contaminated by countless tonnes of dead fish. We will also fight against the degradation of nature on the banks of the river.[17]

In Maria de Fátima Silva's statement, her indignation and that of the other stallholders she represents is clear in relation to the social and economic impact that the construction of the power station has had on the stallholders who depend on the Tocantins River's flow period, usually from June to August, when the beaches are free for the stallholders to set up their stalls and turn their economies around.

The economic activities carried out by the stallholders from the cities of Estreito and Aguiamópolis, Maranhão and Tocantins, respectively, took place every year on the beaches at the foot of the Juscelino

---

[16] Oral report made in the field notebook on 7 November 2012.

[17] Through this letter of Motion of Condolence, the barraqueiro Maria de Fátima Silva, President of the barraqueiros affected by the Estreito HPP, demonstrates that the barraqueiro has a social and economic identity built up over more than 20 years on the banks of the Tocantins River, on the beaches at the foot of the JK bridge and on Cabral Island. Document provided by Maria de Fátima Silva on 09 November 2012.

Kubitschek bridge and on Cabral Island (Annex D). As for the contamination of the beaches that Silva mentions in his letter of Motion of Condolence, this occurred, according to the fishermen of Colony Z-35 in the city of Estreito, shortly after the inauguration of the first turbine of the hydroelectric power plant in 2010. As Costa argues,

> With the activation of the dam's turbine, thousands of fish appeared dead, fish that were part of the fishermen's diet and are now dead because of the construction of this hydroelectric dam that will bring wealth only to those who are already rich. For us fishermen, it's a lot of damage that nobody is going to pay us for. To CESTE,

> we are not entitled to compensation for economic losses because the Tocantins River is not will dry out.

For João Pereira da Costa, a fisherman for over 30 years and a member of the Z-35 Colony in Estreito, the inauguration of this turbine led to the death of thousands of fish (Annex E), which were the fishermen's main source of income and their main diet. According to Vainer (2011, p. 3), "the amount of dead fish reaches 20 tonnes, and the situation is leaving the population of Estreito and the municipalities affected by the project terrified".

In the view of another fisherman, Osmar Rodrigues dos Santos, affiliated to the Estreito fishermen's colony, and with more than 28 years of fishing on the Tocantins River:

> The fish's natural reproductive cycle has become unbalanced due to the construction of the power station itself and also because the Estreito Energia Consortium has not yet built a fish ladder, which means that the fish on the top side of the dam stay there and those on the bottom side only stay at the bottom, not going up[18][19][20] .

According to Osmar Rodrigues dos Santos, in addition to professional and subsistence fishing, the fishermen also developed their vazantes (Annex C) in conjunction with the fishing activity, which made it possible for them to earn a living. For them, there was no compensation either for fishing, their main economic activity, or for the plantations, secondary activities they carried out on the banks of the Tocantins River, which also corroborates the levels of dissatisfaction in this community, as we have tried to demonstrate through the figures and tables in this research.

Evaluating the heterogeneity of the communities surveyed, it was possible to see that, with the exception of traders, whose monetary gain is controlled by the force of the market governed by the law of supply and demand, 58% of all the other communities analysed, in their quest for compensation, had placed their cases in the hands of a lawyer. Of these, 94 per cent preferred to take their claims to a single lawyer; others had already taken their cases to more than one, claiming that the first had been bought by the Estreito

---

[18] Oral report made in the field notebook on 5 November 2012.

[19] Oral report made in the field notebook on 22 November 2012.

[20] Most of the time, the fishermen did not own the land that bordered the Tocantins River, although they were authorised by the landowner to plant their crops and build their straw houses.

Energia Consortium.

In fact, the construction of the Estreito HPP, by territorialising itself in this region of the Middle Tocantins, with the sole aim of exploiting the Tocantins-Araguaia hydrographic basin to generate hydroelectricity, ended up compromising a tangled socio-economic network that had been maintained for decades by various communities affected by the construction of the plant. According to the testimony of market trader Joaquim Lima Rocha,

> The dam put an end to the livelihoods of Estreito, as many people depended on the Tocantins River to survive. We market traders used to buy our produce from the farmers and vazanteiros who produced it at the top of the dam, where it no longer exists. Now we have to bring in some products from towns in the interior of Maranhão and Tocantins, which is very expensive for us.[21]

For this market trader from the city of Estreito, the construction of the power station brought enormous economic losses, as almost all the products sold in the municipal market came from plantations or livestock grown on the banks of the Tocantins River, which allowed for a kind of socio-economic network between those who produced, transported, commercialised and consumed locally and regionally, as shown below.

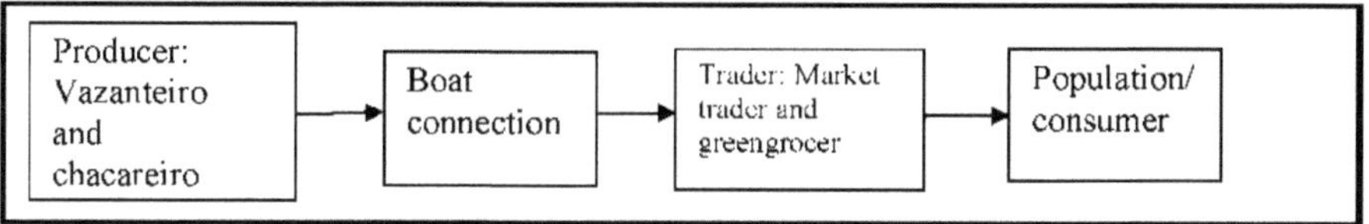

Figure 11 - Circulation of the local-regional economy before the construction of the Estreito hydroelectric power station.
Source: Own elaboration based on field notes in 2012.

This investigation shows that only two of the six communities analysed in this research received compensation for the socio-economic impacts they suffered as a result of the Estreito hydroelectric plant. In this case, however particularised the compensation process may have been, it is not possible to say the same thing about the countless losses, not only social and economic, but also cultural, territorial and environmental, that the communities made up of fishermen, boatmen, small traders and vazanteiros, analysed in this research, suffered as a result of the territorialisation of this engineering project.

---

[21] Oral report made in the field notebook on 15 November 2012.

CHAPTER 4

FINAL CONSIDERATIONS

This research sought to contribute to the debates that revolve around the riverside communities affected by the Estreito hydroelectric power station in the state of Maranhão, focusing on the socio-economic impacts and the socio-political process of the compensation that these communities received or did not receive as a result of the material and immaterial losses they suffered as a result of the territorialisation of this major undertaking.

The main controversy surrounding the debates involving communities affected by the formation of hydroelectric plant reservoirs concerns the great difficulties encountered by the companies building such projects in not knowing how to properly conceptualise who the communities affected by the construction of hydroelectric plants are and/or once they have been conceptualised, what type of compensation should be paid.

In the case of the Estreito hydroelectric plant, the focus of this research, since the preparation of the Environmental Impact Report (EIA/RIMA), finalised in 2002, a series of problems had already been observed regarding the areas subject to flooding and, consequently, possible compensation. In other words, at first, the quota stipulated by the study was 158m, which excluded some municipalities in Tocantins below this quota from possible compensation. Subsequently, this quota was changed and maintained at 156m, which reduced the area that would be submerged by 10.8 per cent, which "[...] was equivalent to a cut in generation capacity of approximately 70 km^2 , leaving the plant with an estimated capacity of 1,050 MW" (ENVIRONMENTAL IMPACT REPORT, 2002, p.9).

While carrying out the field research for this study and analysing and interpreting the data, it was noted that the hypothesis once formulated, based on the social, economic, political and cultural capital defended by Bourdieu (2000) as determining factors of an individual's position in society, was partially confirmed in this case study. The reasons for this lie in the fact that the communities affected by the Estreito power plant are concentrated in a region where cultural (educational) capital has not been a preponderant element, since, in general, the average number of people affected has been around 7.9 years of schooling, which places them in a position of incomplete primary education, theoretically low cultural capital.

Another element that puts the communities analysed in this study at a disadvantage in terms of their relationship with other sectors of society is their low level of involvement in party politics, given that of the 30 impacted people analysed in this study, just over 16% belong to a political party, and none of them have ever run for elected office. This political capital, in the case of this region, was partially confirmed by the fact that these communities are involved in social movements based on demands, which shows in practical terms that they are somehow aware that political capital is crucial to achieving their ideals as social and political actors organised around common objectives.

By basing their compensation policy on a territorial-patrimonialist conception, a policy that has been part of the Brazilian reality since the construction of the first large hydroelectric dams, which, as already mentioned, takes us back to the 1970s and 80s, the construction companies excluded, by classification, the non-owners and a series of other communities that had structured their ways of life on the banks of large rivers such as the Tocantins, among others.

According to Benincá (2011, p. 45), when a hydroelectric plant lays its exploratory foundations for power generation, the entire socio-environmental environment tends to be disrupted as a result: "[...] whether it's the school teacher who lost her job as a result of the project, or the milkman who used to produce his milk in the flooded area and now no longer has a job either," in short, these are individuals who, even though they are part of the territory that will be submerged by the plant's lake, are not compensated for their material losses.

Practically all the political, social, cultural and economic capitals mentioned in this research, defended by Bourdieu (2000) as the elements that define the position of individuals in the contexts in which they are immersed, influenced the struggles for compensation between the riverside communities and CESTE, although obviously the weight that each one exerted in this specific context was differentiated. However, only economic capital and partially political capital were confirmed as determining elements in the compensation received by the communities affected by the Estreito hydroelectric plant. Only farmers with land titles, and to a lesser extent boatmen who were regularly affiliated to the Estreito Boatmen's Association of the state of Maranhão (ABEMA), received compensation for economic losses.

What we realised during this investigation is that, with the exception of the farmers and almost all the boatmen, the other riverside communities are fighting and reluctantly using established legal means in an attempt to sensitise the competent authorities to the recognition of their material rights, which have been totally and/or partially compromised as a result of the construction of the Estreito hydroelectric power station, the dimensions of whose real impacts on the populations that depended on the Tocantins River had not been resolved by the time this research was carried out between 2011 and 2012.

In order to reverse this reality and the countless injustices that have already been committed to thousands of communities affected by the construction of hydroelectric dams, the state, as an intermediary between social classes, should be present at the time of the socio-political process of compensation, considering that, to date, it is the construction companies that dictate the type of impact and the compensation to be paid.

In this way, the companies that build the hydroelectric plants should include, in proportion to the impacts that the riverside communities will suffer, the improvements they have and the link established with the river that will be impacted, an indemnity that includes, among the groups affected and deserving of compensation, not only the landowners, but, in general, all the communities that physically, socially, economically and culturally suffer from the impacts produced by large-scale hydroelectric projects.

CHAPTER 5

REFERENCES

ABRAMOVAY, R. Between God and the devil: markets and human interaction in the social sciences. Revista de Sociologia da USP, v. 16, n° 2, 2004.

NATIONAL WATER AGENCY. Available at: <www.ana.gov.br>. Accessed on: 22 May 2013.

ALMEIDA, G.R. Indigenous lands and the environmental licensing of the Estreito hydroelectric plant: an ethnographic analysis of a socio-environmental conflict. Dissertation (Master's in Social Anthropology) - Postgraduate Programme in Social Anthropology, University of Brasília, 2007. p. 122.

BENINCÁ, D. Energia e cidadania - the struggle of those affected by dams. São Paulo: Cortez, 2011.

BOURDIEU, P.O poder simbólico. Rio de Janeiro: Bertrand Brasil, 2000.

BRANCO, S. M. 1930: O desafio amazônico. 3 ed. São Paulo: Moderna, 2004.

CAMARA, A. A. F. The deterritorialisation of those affected by dams: a case study of the Simplício hydroelectric scheme and socio-environmental damage (RJ/MG). In: XXIX NATIONAL SEMINAR ON LARGE DAMS. Porto de Galinhas- PE, 2013.

CANO, Wilson. Regional imbalances and industrial concentration in Brazil: 1930-1970. São Paulo: Global; Editora da Universidade Estadual de Campinas, 1985.

CASTRO, B.L. Socio-environmental criteria for loss replacement and relocation for those affected by dams: a study of the village of Palmatuba/TO. Dissertation (Master's in Geography) - Postgraduate Programme in Geography, University of Brasília, 2009. 145p.

CHAVES, P. R. Socio-territorial relations in the construction of the Estreito hydroelectric plant and the reproduction of urban space in the cities of Carolina-MA and Filadélfia-TO. 2009. Dissertation (Master's in Regional Development) - Postgraduate Programme in Regional Development, Federal University of Tocantins, UFT/ Palmas- TO, 2009. 198 p.

The World Commission on Dams.Taking on dams and development: a new model for decision-making. Report of the World Commission on Dams. Nov. 2000. Available at: <www.dams.org>.

DUARTE, R. Pesquisa qualitativa: Reflexões sobre o trabalho de campo. Qualitative fieldwork-ethnographic research - research methodology. Pontifical Catholic University of Rio de Janeiro. Cadernos de pesquisa, n. 115, March/2002. p. 139- 154.

FOSCHIERA, A. A. Affected by the Barra Grande hydroelectric plant. Centre for Latin American Policy Studies. In. IV SYMPOSIUM ON SOCIAL STRUGGLES IN LATIN AMERICA. Londrina-PR, September 2010.

GOMES, K. D. Socio-economic characterisation and perception of fishermen in the Tocantins River immediately downstream of the Lajeado HPP dam. Palmas, 2007.

GONÇALVES, C, W. P. The (un)paths of the environment. São Paulo: Ed. Contexto, 2006.

GRANOVETTER, M. Economic action and social structure: the problem of immersion. FORUM OF ECONOMIC SOCIOLOGY. 2007. Available at: <www.rae.eom.br/eletr0niea>. Accessed on: 20 January 2012.

GUJARATI, D. N. Basic Econometrics. 3 ed. Rio de Janeiro: Elviser/Campus, 2006.

BRAZILIAN INSTITUTE OF GEOGRAPHY AND STATISTICS. Hydrographic Basins. Available at: <www.bacia hidrográfica. ibge.gov.br>. 2000. Accessed on: 12 May 2013. I
BRAZILIAN INSTITUTE OF GEOGRAPHY AND STATISTICS. Demographic census, 2010. Available at: <www.censo 2010. ibge.gov.br>. Accessed on: 25 May 2013.

LAMONTAGNE, A. Os impactos do processo de licenciamento ambiental: análise da administração estatal do conflito socioambiental, interétnico e multicultural da usina hidrelétrica de Estreito. Dissertation (Master's in Comparative Studies on the Americas) - Postgraduate Programme in Comparative Studies on the Americas, University of Brasilia, 2010. p. 148.

LIRA, Elizeu Ribeiro. *The* genesis of Palmas-Tocantins: The geopolitics of territorial (re)organisation in the Legal Amazon. Goiânia: Ed. Kelps, 2011.

MAB. National Movement of People Affected by Dams. Available at: <www.mabnacional.org.br>. Accessed on: 23 May 2012.
MATIELLO, C. Practices and representations of the military dictatorship in Itaipu binational propaganda. In: XXIII NATIONAL SYMPOSIUM ON HISTORY. Londrina, 2005.

MELO N.L; CHAVES, P.R. The construction of the Estreito hydroelectric plant and the territorialisation process of the Movement of People Affected by Dams. MG. In: XXI NATIONAL MEETING ON AGRICULTURAL GEOGRAPHY. Uberlândia - MG, 2012.

MINISTRY OF PLANNING. Budget and Management, 2010. Available at: <www.Planejamento.gov.br>. Accessed on: 04 May 2013.

FEDERAL PUBLIC MINISTRY. Public Civil Action, 2011. Available at: <http://www.mpf.mp.br/>. Accessed on: 2 May 2013.

MITCHELL, R. C; CARSON, R. T. Using Surveys to Value Public Goods: The contingent valuation method. Resources for the future. Washington. D.C, 1989.

ENVIRONMENTAL IMPACT REPORT (RIMA). Environmental Impact Assessment (EIA) of the Estreito Hydroelectric Power Station, located on the middle stretch of the Tocantins River, Maranhão, CNEC, 2002. 83p.

REPÓRTER BRASIL. News, communication, research and journalism agency. Labour rights and socio-environmental damage. Available at: <www.repórter brasil.org.br>. Accessed on: 02 March 2013.

REPÓRTER BRASIL. News, communication, research and journalism agency. Installation and operationalisation of the Estreito power plant (2007). Available at: <www.repórter brasil.org.br>. Accessed on: 02 March 2013.

RIDENTI, M. Política pra quê? São Paulo: Atual, 1992 (Living history series).

SCARLATO, F. C; PONTIN, J. A.O Ambiente Urbano. São Paulo: Atual, 1999.

SEVÁ FILHO, O. Strange cathedrals. Notes on hydroelectric capital, nature and society. In: Science and Culture Magazine. SBPC, July/September 2008.

SIEBEN, A. State and energy policy: the deterritorialisation of the rural community of Palmatuba (TO) by the Estreito hydroelectric plant. Thesis (PhD in Geography) - Postgraduate Programme in Geography, Institute of Geography, Federal University of Uberlândia, 2012. 204 p.

SILVA, S.G. R; SILVA, P.V. Those affected by dams: reflections and theoretical discussions and those affected in the Olhos D' Água settlement in Uberlândia-MG. Society and Nature Magazine. Year 23; n.3, p. 397-409, Sep/Dec.2011.

SKIDMORE, T. Brasil: de Getúlio Vargas a Castelo Branco. Rio de Janeiro: Paz e Terra, 1976.

THOMPSON, E. P. 1998. The moral economy of the English mob in the eighteenth century. In: . Customs in common: studies in traditional popular culture. São Paulo:
Companhia das letras, p. 150-202.

VAINER, C. B. Population, environment and social conflict in the construction of hydroelectric dams. Coletânea momento. In: MARTINS, George (Org.) Population, environment and development: truths and contradictions. 2 ed. Campinas - SP, 1996.

VAINER, C. B. Why are they dying? Tons of dead fish in the Estreito - MA hydroelectric dam lake worry the population and environmentalists. *Jornal Momento do Maranhão.* 7 Apr. 2011. p. 3.

WALDMAN, M. Ecologia e Lutas Sociais no Brasil.6 ed. São Paulo: Contexto, 2002.

ZITZKE, V. A. *The* socio-technical network of the Lajeado hydroelectric plant (TO) and the rural resettlements of the families affected. Thesis (PhD in Human Sciences) - Postgraduate Programme in Human Sciences, Federal University of Santa Catarina. Florianópolis, Apr. 2007.

CHAPTER 6

ANNEXES
ANNEX A - Concentration of Brazilian industrial activity on a regional scale South-Southeast - 1970

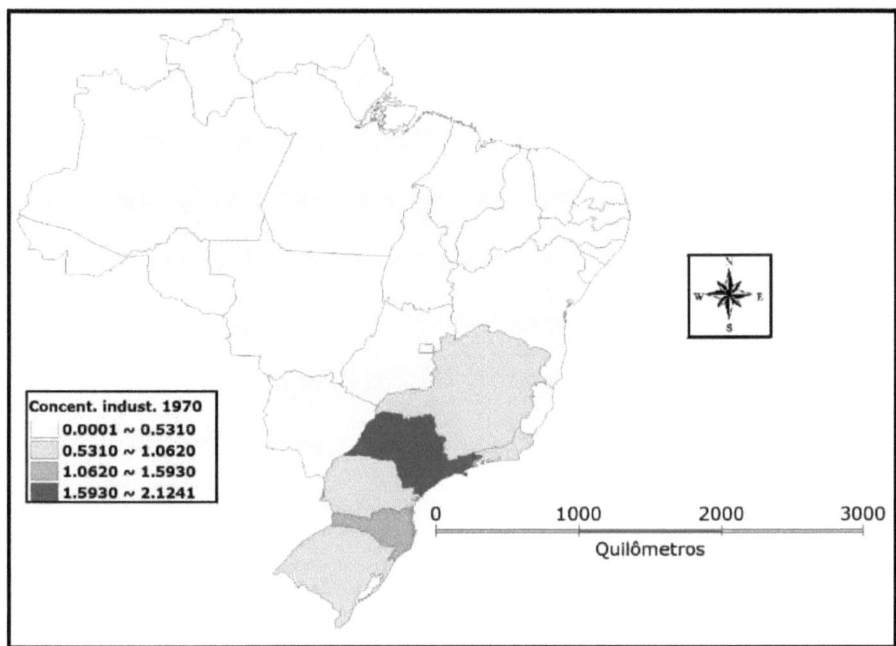

Source: Pereira, M. based on Cano.

ANNEX B - Brazil's main river basins up to the year 2000

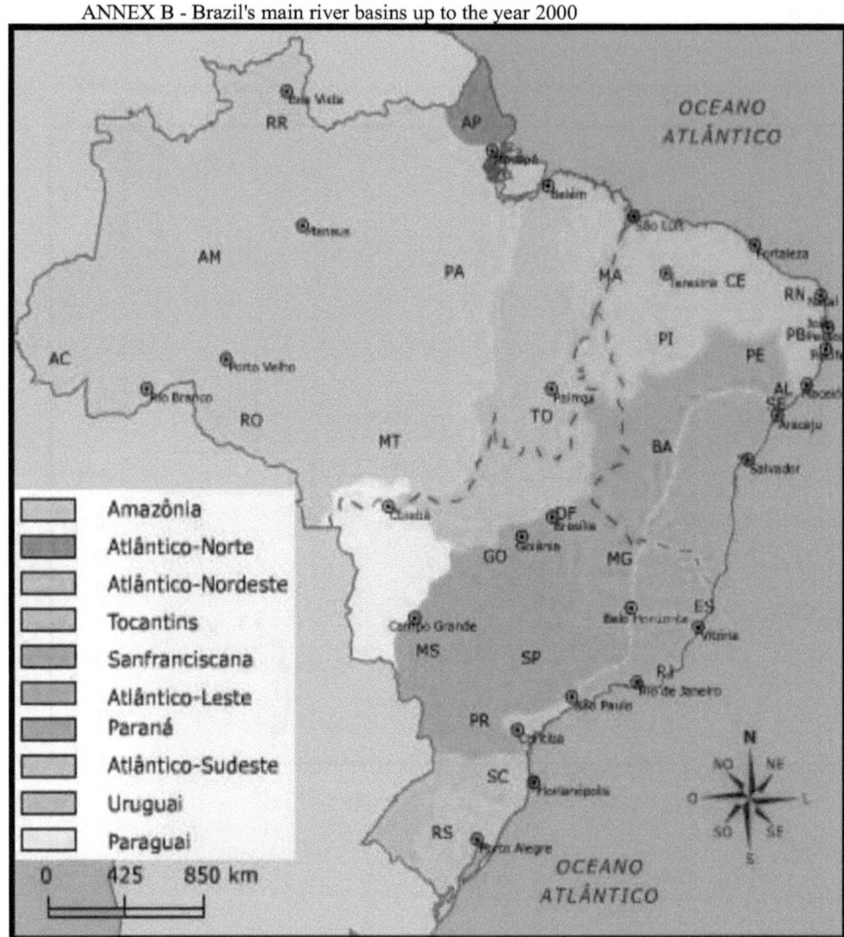

Source: IBGE -2000

ANNEX C - Primary activities on the banks of the Tocantins River before the construction of the Estreito dam

Source: Pereira, M/ November 2012.

Source: Pereira, M/ November 2012.

ANNEX D - Beaches at the foot of the JK bridge and Cabral Island, where the stallholders carried out their economic activities before the power station was built

Source: Pereira, M/ November 2012.

Source: Pereira, M/ July 2011.

Source: Pereira, M/November 2012.

Source: Pereira, M/November 2011.

ANNEX F - Products sold in the Estreito municipal market before the dam was built, grown on the banks of the Tocantins River.

Source: Pereira, M/November 2012.

Source: Pereira, M/ November 2012.

Source: Pereira, M/November 2012.

Source: Pereira, M/November 2012.

QUESTIONNAIRE

Personal details

Name: ____
Year of birth: ________________//
Place of birth: State/city/

Level of education|

(D
( ) didn't study
( ) Iº grade incomplete
( ) Iº completed high school
( ) 2º grade incomplete
( ) 2º completed high school
( ) Superior

Marital status|

(2)
( ) Single
( ) Married
( ) Separated
( )Friend
( ) Widowed
( ) Other

(3) How many people live in your house? Who are they?
4)   What was/is your father's occupation?
5)   5) What was/is your father's schooling?

6)   6) What was/is your mother's schooling?
7)   7) What economic activity/occupation did you carry out before and after the dam?
8)   8) What kind of improvement(s) did/do you have?
9)   9) Do the activities you developed and are developing near the Tocantins River take place on your own property? Leased property? Rented? For how long?
10) 10) Was your link with the river permanent? Temporary? What was your main activity? For how long?
11)   Have you or any of your relatives ever joined a political party? Do you hold elective political office? Do you hold any non-elective political office? Which one(s)? When?

Indemnity process

12)   With the construction of the Estreito HPP, in the middle of Tocantins, what kind of losses did you suffer as a result of this development?

(13)Have you appealed to any body or legal representative for possible compensation for material loss as a result of the construction of the Estreito HPP?
( ) Yes.
( ) No.

If so, which one/how?

If not, why?

(14) Do you receive any kind of help as a result of the losses you mentioned above?
(15) Receives some kind of government aid
( ) yes ( ) no
What/How/When/Whom?

(16) Before the construction of the Estreito HPP, and considering your main economic activity, what was your family's average income?

(17) After the dam, did your family income increase or decrease? How much more or less can you earn today?

Destination of production

(18)   Considering your occupation and the main economic activity you develop or developed near the Tocantins River, what was/is the destination of the production?

Basic infrastructure before and after the dam

(19)Electrical network

                        BeforeAfter
( ) yes ( ) no yes ( ) ( ) no

Piped water before
( ) public network ( ) artesian well ( ) simple well ( ) mine ( ) other

Piped water after
( ) public network ( ) artesian well ( ) simple well ( ) mine ( ) other

Water treatment used before
( ) filtered ( ) chlorinated ( ) none.

Water treatment used after
( ) filtered ( ) chlorinated ( ) none.

Socio-political representations

(20)   Were you part of any social movements before or after the dam was built? What was it? Why?
(21)   Were you or are you part of any associations or co-operatives set up before? During? Or even after the dam was completed?

(22)

yes
I want morebooks!

Buy your books fast and straightforward online - at one of world's fastest growing online book stores! Environmentally sound due to Print-on-Demand technologies.

Buy your books online at
www.morebooks.shop

Kaufen Sie Ihre Bücher schnell und unkompliziert online – auf einer der am schnellsten wachsenden Buchhandelsplattformen weltweit! Dank Print-On-Demand umwelt- und ressourcenschonend produziert.

Bücher schneller online kaufen
www.morebooks.shop

info@omniscriptum.com
www.omniscriptum.com

Printed by Books on Demand GmbH, Norderstedt / Germany